Field Guide to

OLD-GROWTH FORESTS

Field Guide to

OLD-GROWTH FORESTS

Exploring Ancient Forest Ecosystems from California to the Pacific Northwest

Text and Illustrations by Larry Eifert

SASQUATCH BOOKS
SEATTLE

Printed in the United States of America
Distributed in Canada by Raincoast Books, Ltd.
06 05 04 03 02 01 00 5 4 3 2 1

Cover design and illustration: Dugald Stermer
Interior design: Kate Basart
Interior illustrations/photographs/maps: Larry Eifert

Library of Congress Cataloging in Publication Data
Eifert, Larry.
Field guide to old-growth forests : Exploring ancient forest ecosystems from California to the Pacific Northwest / text and illustrations by Larry Eifert.
p. cm.
ISBN 1-57061-234-X (alk. paper)
Old growth forest—Northwest, Pacific—Guidebooks. 2. Old growth forest ecology—Northwest, Pacific—Guidebooks. 3. Forest plants—Northwest, Pacific—Identification. 4. Forest animals—Northwest, Pacific—Identification. I. Title.

QH104.5.N6 E46 2000
578.73'09795—dc21 99-057284

Sasquatch Books
615 Second Avenue
Seattle, Washington 98104
(206) 467-4300
www.SasquatchBooks.com
books@SasquatchBooks.com

Other titles in the Sasquatch Field Guide series:

Adopt-a-Stream Foundation
Field Guide to the Pacific Salmon

The Audubon Society
Field Guide to the Bald Eagle

Great Bear Foundation
Field Guide to the Grizzly Bear

The Oceanic Society
Field Guide to the Gray Whale

American Cetacean Society
Field Guide to the Orca

People for Puget Sound
Field Guide to the Geoduck

The Western Society of Malacologists
Field Guide to the Slug

Greater Yellowstone Coalition
Field Guide to the North American Bison

Contents

Acknowledgments

Nature was of the utmost importance to my family; it was more than the fact that both of my parents made their livings teaching and interpreting the great outdoors. We enjoyed all of nature: the forests and prairies, the marshes and the rivers, even roadsides left shaggy by highway department mowers. In my family there was no such thing as a weed; rather, it was just an out-of-place plant needing a proper home. Even though we were thousands of miles from them, old-growth forests of the Pacific Northwest held the highest fascination for us. We thought of those coastal fogscrapers as almost sacred, the ultimate expression of what trees should be. As a child, I never saw these old-growth trees, but I knew of their existence and that was enough for me.

Most of these trees I took for granted thirty years ago are gone today. Less than 10 percent of the original presettlement ancient forests remain in British Columbia, Oregon, and Washington, and less than 4 percent of the redwoods in California still reach for the sky.

Great thanks, for embedding in me a lifelong passion for forests, goes to prolific author and photographer Virginia Eifert and to ecologist Herman Eifert, without whom "nature" wouldn't have created this writer and painter, without whom this field guide wouldn't exist. I would also like to thank my friends at California State Parks, the National Park Service, and Save-the-Redwoods League for their continued support of my tree projects. Thanks as well goes to my partner and wife, Nancy, for pushing me further up the path of nature interpretation each year. I hope the journey never ends.

Streams are important components of old-growth forests

Introduction to Old-Growth Forests

In the length of one lifetime the world-class Pacific Northwest natural heritage has almost disappeared. Where there were once millions of the tallest and most impressive trees on earth, now there are only small little pockets of remote forest. Not many years ago, lowland ancient forests were common and accessible; these days, however, most people do not know where or even what an old-growth forest is. We bring out the map, stab abstractly at various tiny green spots and say something about remote, difficult-to-find areas that are well worth the hunt once one gets there. Do *you* know where to walk beneath old growth? Where to experience the wonder and awe of thirty-five-story trees or thirty-five kinds of moss growing in one locale?

This field guide is not about weighty scientific details, or the disputes about or a denunciation of past timber practices. Instead, it's about access and appreciation. Driving along the road, one sees countless trees growing, but most of these are little guys, young and unimposing. Forests are everywhere in the Northwest, but these endless rows of fiber crops are not old-growth forests. This book is about the remaining timber ancients, the enduring legacy that humans have almost destroyed.

The old-growth enclaves that still exist are priceless, far more valuable as scientific and spiritual commodities than what the high-quality

Flitty denizen of old-growth forests—the chestnut-backed chickadee

close-grained wood is worth as decking or hot-tub lumber. These unaltered landscapes are critical to the creatures and plants that are parts of the forests. Education is empowering, and knowing more about these extraordinary places may help preserve some of them.

FINDING THE CARBON RIVER FOREST

The directions were vague, the project was to create a large wall mural for Mount Rainier National Park, and the theme was ancient forests. I had heard of the Carbon River Forest before. Located on the lower northwestern flank of that immense volcano called Mount Rainier, the ice slopes of the Carbon Glacier seem to keep the valley below them cooler and wetter than others, and the mountain's bulk increases precipitation. Over thousands of years these conditions have produced a lush lowland rain forest. This land has been under the auspices of the National Park Service for more than a hundred years; it therefore remains a relatively unaltered enclave of big trees.

I drove through the urban confusion of Tacoma and Puyallup, past rows of auto dealerships and strip malls stretching endlessly along the asphalt until we found the little town of Buckley. A stump 12½ feet in diameter, in the town park affirms that this area must have at one time been big forest, but it is difficult to imagine. Other trees are in sight, but smaller and oddly shaped, like in most other Puget Sound communities.

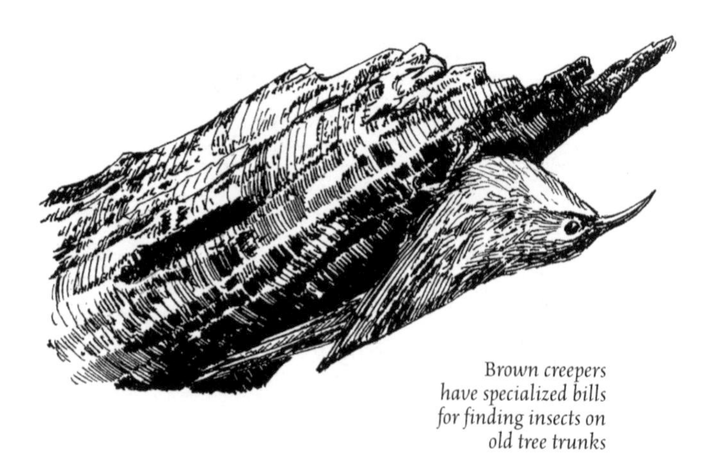

Brown creepers have specialized bills for finding insects on old tree trunks

South of Buckley the road enters coal country, passing remnants of old mining towns with such names as Melmont, Manley-Moore, and Montezuma. Rows of uniform little trees cover most hillsides; freshly logged clearcuts cover other hills, huge scars with no visible vegetation in the process of being replanted with another fiber crop. Here was a patchwork quilt of foothills, some green, others brown, with magenta fireweed and foxglove in bloom on many slopes. This was a landscape being worked hard. In the flatlands of the Midwest, these would be corn and soybean farms. Across the old Fairfax Bridge, a rickety affair guiding me over the deep Carbon River gorge, the road became more remote. I soon passed a junction to Mowich Lake. With every mile ever-steeper hills became even more ragged and torn up. Logging slash was piled high; freshly burned clearcuts everywhere.

THE END OF THE ROAD

As I came closer to what appeared to be the end of the road, I saw that the road didn't really end at all. Instead, it disappeared into the forest, swallowed up by a wall of trees. The park boundary was distinct: a knife-straight line of trees separating the uncut from the cut, the national park's land from national forest. Past the park entrance station I slowed the car down to a crawl. With windows rolled down and my head out, I gaped at the view. I had entered a different world. The contrast between "outside" and "inside" that forest, under the canopy, was like the opposites of high noon and dusk, hot and cool, noisy and quiet, ragged and lush.

I pulled into a little turnout. Sounds of gurgling water from a small culvert lured us out of the car. There in the half-light and slanting shadows of this seasonal streambed was a mural just waiting to be painted. Here was that classic old-growth scene I had been searching for. I dropped down to the stream gravel and a riot of plants greeted me. Above winter's high-water line, moss covered almost every inch of surface. Ferns, flowers, and small shrubs grew above the moss, and countless numbers of young trees created nurseries along each downed log. Flowers crept out of the moss carpeting; more moss hung from the enormous trees and shrubs overhanging the stream. No space seemed to be unused; much was festooned by more than one plant. It was a crowded place, yet absolutely nothing seemed out of place.

The overall impression was hypnotic. My senses seemed to heighten; harmony and regularity were the overwhelming sensations. One is *on* a prairie or beach, but *in* an old-growth forest, surrounded on all sides by it, overwhelmed by it. Compared with the trees, I seemed small, but to the mosses and tiny flowers I was a giant. I felt uncivilized in the forest stillness, even my breathing seemed too loud. To my nose, the subtle fragrances of dry red cedar and wet moss were pure perfume. I felt at once like an alien but so at home I never wanted to leave. A winter wren chattered in the background.

The small stream was also interesting: As transparent as glass, it was a golden ocher. As the stream meandered around an enormous hemlock, the tree's giant root system created a waterfall. Below this, small pools reflected lime green as salmonberry leaves overhung the riffles. Well-armed spikes of devil's club hugged the far shore, its red blackjack of flowers daring a closer look. Nearer, at about chest height, a huge log lay completely hidden by mosses and other tiny plants. At least I assumed there was a log underneath; not an inch of actual wood could be seen. How long ago had this great tree toppled? How many countless living things had begun life aboard this nutrient storehouse, this stationary forest ark?

Stooping slightly to bring my eyes to log level, I gazed across the mossy top to carefully inspect a miniature forest of false lily-of-the-valley, starflowers, and bunchberry. Here was a perfectly manicured garden of dainty white flowers. A tree frog sat motionless, its tiny eyes as camouflaged as its mottled green skin. Carefully pulling back a section of spongy moss, I became entranced by a thick network of whitish fungal hairs, roots, and intertwined tunnels—highways for beetles, mites, termites, carpenter ants, and a thousand other creatures. The liquid flute-like song of a Swainson's thrush brought me to my senses. A few yards downstream, another small pool sparkled, this one created by a log that completely blocked the stream's flow. Golden sand and bronze gravel reflected good salmon spawning grounds in the cool water. An ageless yet vibrant aroma of the old-growth forest filled my senses as a bit of breeze tossed leaves about.

I drew sketches, took reference photos, and the mural was eventually installed in the Ohanapecosh Visitor Center. Countless visitors now learn the secrets of old-growth forests from this project. The real satisfaction of this experience, however, was the adventure of learning about this unique bit of forest. I will always remember standing quietly beside that little unnamed stream, hearing a symphony of living water, winter wrens, and chickadees. I will always recall the salamander maneuvering along its log home, the sunbeams and shadows passing across an ancient tree's mossy bark, the shimmer and reflection of lime and chartreuse in the golden water.

As I returned to the world of tiny trees and bruised earth just outside the old-growth forest, I felt an urgency to explain what I had just seen and experienced. Forests such as the Carbon River once covered the Northwest. Today they are a scarce bit of our cultural and spiritual heritage; to some they are as meaningful as churches and synagogues. They are the scant remains of what was once the greatest forest on earth.

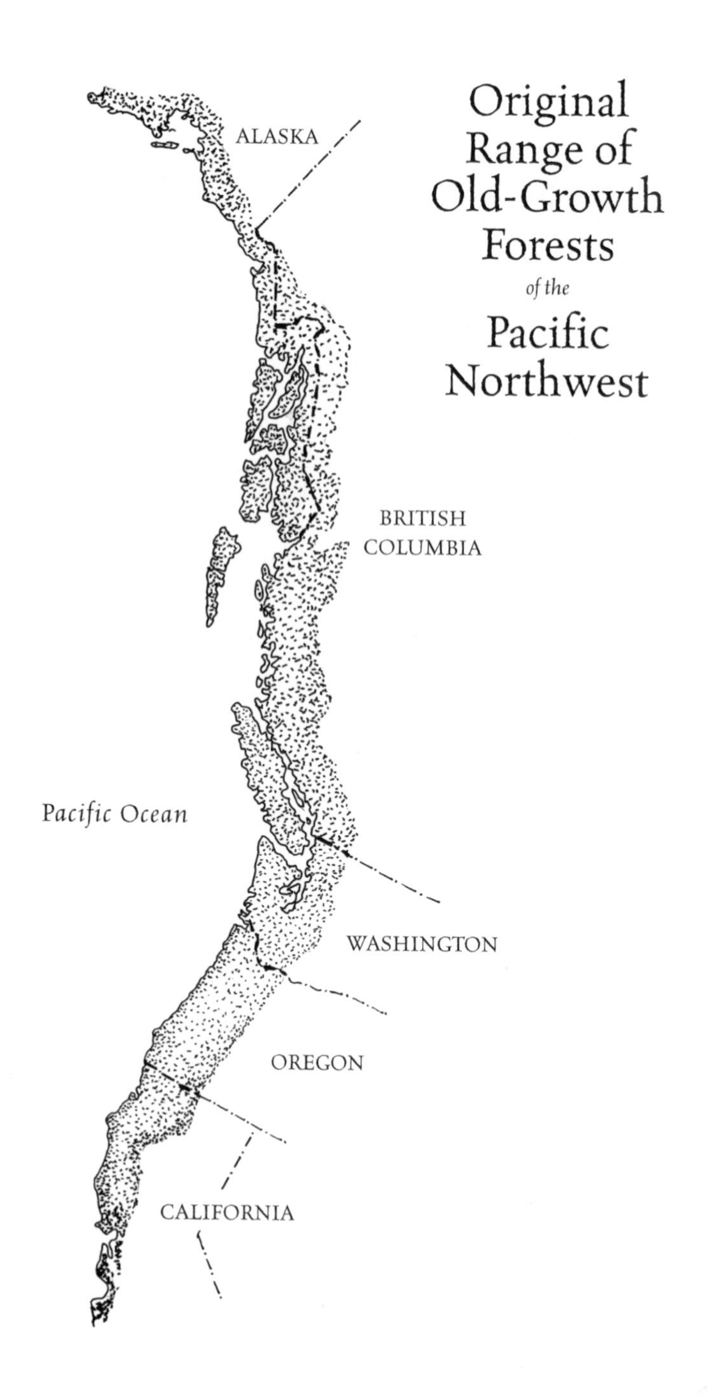
Original Range of Old-Growth Forests of the Pacific Northwest
ALASKA
BRITISH COLUMBIA
Pacific Ocean
WASHINGTON
OREGON
CALIFORNIA

Fog helps maintain the last California coast redwoods

Lowland Old-Growth Forests of the Pacific Northwest

What is old growth? Simply stated, old-growth or ancient forests are defined as forestlands that have not been altered by humankind. Large and unique trees of great age dominate the scene in such forests, while several dispersed layers of lesser trees and shrubs grow beneath them. Disruptions occur to these ecosystems in the form of fire, flood, or windstorm, creating temporary openings that fill with various sun-loving plants. Given enough time, these openings will return to the original constant and stable condition. Logs that alter streamflows dominate watercourses in ancient forests and the forest canopy shields water from the sun, so erosion is slight and changes over time are modest. Over thousands of years, these original forests have reached stability, although plant competition continues to slowly alter some forest characteristics.

When Europeans first came to the northwestern edge of North America, the most impressive forest on earth stood as a barrier. In 1828, explorer Jedediah Smith crossed northwestern California toward the coast and headed north. In journals he commented on the spruce, fir, and redwood, the latter being the noblest trees he had ever seen. The journey was nightmarish, with his party fighting thickets and underbrush, soaking rains, and fog; sometimes they progressed only a mile or so in a day. In what became California's Del Norte Redwoods State Park in 1925, the bible-toting Smith aptly christened the area "Damnation Creek."

For years these forests were not well understood. Scientific studies were late in coming, and professional foresters naively referred to these highly evolved natural communities as "overmature," decadent, or as biological deserts in need of a good cleaning and cutting. And trees were certainly cut. Today, almost the entire original forest heritage is gone, replaced by commercial fiber farms, cities, and urban areas Northwest residents call home. It is not difficult to understand why the tall trees were cut, for they were among the world's finest sources of timber. For a century, they fueled the Pacific Northwest's economic engine. It is difficult to acknowledge now that this important piece of the planet's rich resources has all but vanished.

How did these unique forests develop in the first place? That story weaves together climate change, evolution, and a past Ice Age in a fascinating tale. Water (and lots of it) makes the Pacific Northwest the conifer capital of the world. In winter moisture-laden winds blow in from the Pacific Ocean, then rise to pass over peaks of the Coast and Cascade

Second growth forests

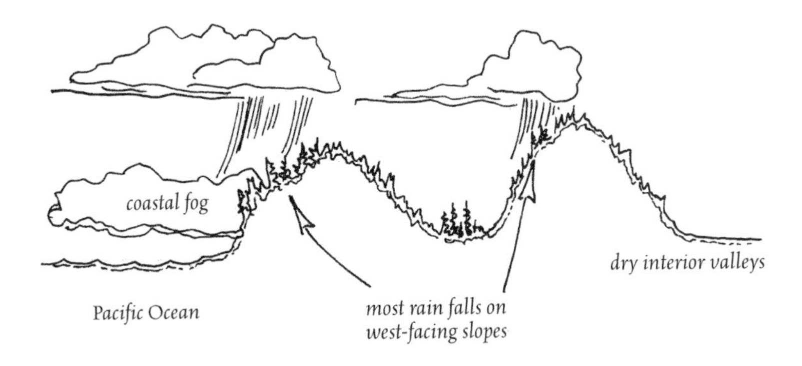

Wet ocean winds rise as they meet coastal mountains, and rain occurs

Mountains. As these wet winds ascend, they drop large amounts of rain and snow. Along the western coastal strip, summer fog adds significantly to this moisture, adding nearly 30 percent to the tally of normal precipitation. Fog is critically important to coastal lowland forests and gives them an essential shot of moisture during the driest summer months. Some local microclimates create truly impressive precipitation totals. For example, although Seattle and Portland both average about 40 inches of precipitation each year, valleys west of Washington's Olympic Mountains, the western coast of Vancouver Island, and the watersheds west of Cape Mendocino, California, commonly receive 125 inches of rain. Some locations even exceed this amount.

All of this water creates forest plants of impressive size and diversity, not the least being the earth's tallest trees. Twenty-five different conifer species live in the Pacific Northwest, with the largest and some of the oldest represented. Seven of these species, coast redwood, Sitka spruce, Douglas-fir, western red cedar, Alaska cedar, Port Orford cedar, and incense-cedar—commonly live longer than five hundred years. Several species reach diameters of 10 feet or more, and two tower 300 feet or higher. It's the complexity of old-growth forests that is most interesting, however, and naturalists are just beginning to fully comprehend how complicated and finely tuned these areas really are.

Old-growth forests can easily be compared with a single living entity, a person, animal or other life-form. Just as in a living body, nothing in an old-growth forest works alone or independently. If so, the entire system becomes altered, out-of-kilter, or sick. If enough parts are removed, the system just ceases to function. Forests are not merely odd collections of trees with a few flowers and birds tossed in; rather, they are interconnected and extremely complex webs of life. Trees are the obvious main structural parts, with needles, leaves, roots, and trunks the evident external shapes, but forests are much more than that. Trees have roots that transport nutrients and hold the trunk upright to the sky. Many fungi, birds, and animals aid tree health; in turn, trees provide other plants and creatures with homes and food. Exactly how forest creatures and plants function in this manner is just beginning to be understood by ecologists. Unfortunately, with so little old-growth remaining, the research may be too late to unravel important secrets.

A Survey of Old-Growth Forest Communities

The landscape of the Cascade Mountains west to the Pacific Ocean knits together several forest communities into three major forest domains. Each forest community is named for the tree species

Coast redwood forests are rich and varied

that grows most abundantly there—its dominant conifer. From south to north these are the forest provinces of the coast redwood, Douglas-fir, and Sitka spruce–western hemlock. Each domain has great similarities, yet each also has interesting and unique differences related to its climate, geography, and associated plants and wildlife. Some plants and animals live in only one community, while others are able to live in all forest domains. These are not separate and distinct forests, however. Rather, they blend along their edges and are sometimes even dependent upon each other.

COAST REDWOOD PROVINCE

Coast redwood forests occupy a slim strip of land just inland from the Pacific coast, from California's Big Sur to southern Oregon. Because they require wet winters and summer coastal and valley fogs, few redwoods grow more than 40 miles east of the ocean. In 1850 redwoods covered about 2 million acres; less than 4 percent remain uncut today. Almost all remaining old-growth redwoods are now protected. Coast redwood forests contain the largest biomass and hold the earth's tallest trees. One redwood may contain enough wood to build fifty average-size homes. Some redwoods are more than 360 feet tall. Battles still rage over the future of the last remaining privately owned trees, as many people continue to express concern for these ancient giants.

DOUGLAS-FIR PROVINCE

Inland from the coastal strip, the Douglas-fir forest stretches from northwestern California to southern British Columbia. It is the only lowland old-growth forest community that commonly withstands summer droughts and requires periodic wildfires for the forest to persist. At one time all the fertile lowland inland valleys of Oregon and Washington held enormous Douglas-firs, some reportedly 400 feet tall, larger than any tree alive today. Douglas-firs are still the most common conifer in the Northwest, but remaining old-growth Douglas-fir is mostly tucked away in remote mountain valleys, where poorer soil and growing conditions create smaller trees. Roughly 10 percent or less of the original Douglas-fir forest remains. Hemlock and red cedar are replacing Douglas-fir on many tree farms. (The Douglas-fir takes its name from the botanist who first identified it around 1827; unlike like it, true firs have cones that rise upright on their branches.)

SITKA SPRUCE–WESTERN HEMLOCK PROVINCE

This is the most northern and wettest of the forest provinces, hugging the coast from Northern California (where it intermingles with the most northern coast redwoods), through the coastal forests of British Columbia, and on into Southeast Alaska. Moderate temperatures with frequent and abundant rainfall mark this region. Cool wet summers with little summer drought and few wildfires provide a stable refuge that grows enormous trees. Found here are the world's largest Sitka spruce, Douglas-fir, western hemlock, western red cedar, subalpine fir, and Alaska cedar. Big-leaf maple, vine maple, and dense layers of moss and epiphytes ("air plants") also define this domain, where forty to fifty species of moss may be found in a single forest. The amount of old growth still standing here is small, however, and big spruce trees are rarer than their coast redwood counterparts.

From the Pacific Ocean to the summits of the Cascade range, there were once about 70,000 square miles of low-elevation old-growth forests in the Pacific Northwest. Today, at least 90 percent have been cut in the United States and slightly less in British Columbia. What is protected is marginal; what remains unprotected will surely fall within the next human generation. Each old-growth tree should be considered irreplaceable. Cut more old-growth—we might as well burn the Smithsonian!

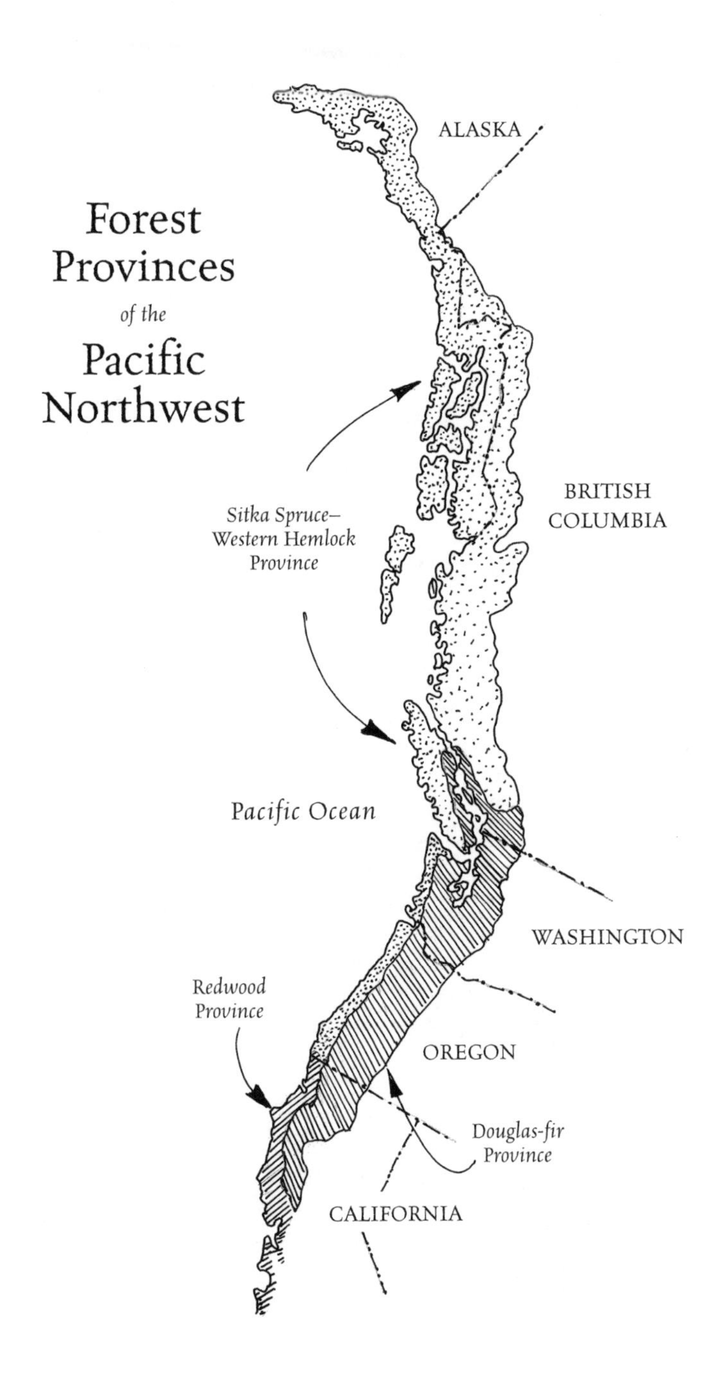
Forest
Provinces
of the
Pacific
Northwest
ALASKA
BRITISH
COLUMBIA
Sitka Spruce–
Western Hemlock
Province
Pacific Ocean
WASHINGTON
Redwood
Province
OREGON
Douglas-fir
Province
CALIFORNIA

Characteristics of an old-growth forest

Anatomy of an Old-Growth Forest

Several basic structural traits create old-growth forests: big trees, dead snags, downed logs, and a multilayered forest canopy. Huge and distinctive trees dominate the scene. These trees begin to acquire an "old-growth" appearance as early as 175 to 200 years of age, but old growth is still considered young at 250 years. Some redwoods grow upwards of 2,000 years; Alaska cedar approaches twice that figure.

TREETOPS

The frontier of forest science today is high in the upper canopy, and what researchers are finding in treetops has them hopping. Little solid research has ever been conducted in what is actually most of an old-growth forest: that area between the tree's lowest and topmost branches. Getting to these research sites has not been easy, for it requires the skills of a mountain climber and a stomach of iron to climb their heights. A research crane near Carson, Washington, now assists researchers.

Life abounds in the treetops

Fallen logs provide important nutrients to future forests

What scientists have discovered in treetops is extraordinary. Fungi and debris from needles build soil up to several feet thick on the big branches, where entire forest communities exist. Voles and salamanders live here, along with owls, woodpeckers, swifts, and murrelets. Huge branches grow 50 to 60 feet out from the trunk and reach for the sky; they are themselves larger than most trees east of the Mississippi River. Lichens, liverworts, fungi, and moss are also found here, thick and lush, and have been found to make up as much as one ton of plant material for every two acres of trees. Such forest communities begin to develop only after many years, and only after the trees have begun to mature.

DEAD STANDING TREES AND FALLEN LOGS

Old-growth trees are ancient indeed, but trees don't live forever. The giant conifers are especially vulnerable to winter storms, parasites, disease, and fire. When a tree dies and becomes what's called a snag, hosts of insects, larvae, spiders, fungi, and microscopic organisms quickly make these places a vibrant column of life, an important ingredient of the entire ecosystem.

As beetles, termites, and carpenter ants attack dead trees, woodpeckers and other birds gather to feast on the decomposers, chiseling deep holes to get at their prey. In turn, this allows increased access for more insects and provides nesting sites for such birds and animals as woodpeckers, owls, and flying squirrels, to name a few. As the scars of fire in existing old-growth areas confirm, wildfires may revisit and burn these snags many times, eventually hollowing out trunks and roots.

Bears climb up snags and den in hollow trunks and rotted-out cavities. Slabs of dead and loosened bark also create ideal nooks for roosting and hibernating bats; broken treetops become nesting platforms for eagles or ospreys. Early settlers were quick to use fire-hollowed trees. Such "goosepen" trees housed poultry, and some early settlers even lived in them. The decay process generates heat, making it cozy in the winter for all cavity dwellers.

Researchers estimate that about one hundred of the Pacific Northwest's forest bird species, two hundred varieties of mosses and lichens and fifteen hundred different invertebrate species use standing snags in various ways. Snags are busy and beautiful places. Old trees take on unique characteristics, with bumps, bruises, and scars, their odd careening branches and leaning trunks attesting to the gnarled legacy of battling the elements in both life and death.

Old trees or snags eventually fall to the ground, taking others down as they go in a jumbled heap of twigs, mud, and debris. The event opens a "skyhole," as sunlight floods a once-darkened forest. This is the opportunity for which many plants have been waiting. Young trees, shrubs, and herbaceous plants race to fill the space, taking advantage of the bright light streaming in. To germinate, Douglas-fir seeds require not only full sun but also disrupted soil, and many groves of even-aged old firs show past forest openings caused by fire. Attracted by this new growth, many creatures come to browse the new landscape. Black-tailed deer and Roosevelt elk may appear, lighter-colored Bewick's wrens displace darker and smaller winter wrens, and Townsend's voles replace the smaller muddy-colored creeping voles. These shifts occur because darker species are better camouflaged in dark forests, and smaller species require less food, both of which are characteristics of dense old-growth forests. Lighter-colored and larger species do well in sunny areas.

After several decades, the skyhole is filled with mature trees; sun-loving shrubs and flowers are shaded out, and initial canopy development is complete. From a distance, a healthy old-growth forest appears as a collage of textures—with large old trees, younger stands, and sunny openings.

MULTILAYERED CANOPY: Many Layers, Many Plants

Unlike even-aged tree farms, where all the trees in a given area were planted at the same time, ancient forests include many different species of shrubs and small trees that create several layers of forest between the treetops and soil. These intermediate layers represent an important manifestation of old-growth forests. Responding to various degrees of sunlight, most of these plants have relatively short lives when compared with ancient conifers, but they all help to create a more varied and complex forest. Some conifers, such as western hemlock and redwood, remain small and shrublike for years, until a skyhole provides an opportunity for them to reach for the sun.

DOWNED LOGS AND LOGS IN STREAMS

Traveling cross-country in an old-growth forest is exasperatingly slow, like an ant trying to navigate through a partially played game of pickup sticks. Logs stack up on each other at odd angles, 15 feet high or more. Sometimes when a tree falls, its huge rootwad upends, excavating an enormous hole. If the falling tree hits a waterway and partially dams it, changes in the stream's flow may be created. These alterations contribute in many ways to the forest's overall health and are even more important to the old-growth ecosystem than standing snags are.

Downed logs may last many hundreds of years before completely decomposing. In some forests as much as 265 tons per acre of downed wood

Many western hemlocks begin life atop nurse logs

The tiny winter wren—an avian "forest mouse" common in old-growth forests

has been measured; even more has been measured in streams, where the lack of oxygen in waterlogged wood slows fiber decay. These downed logs slowly add nutrients to the surrounding forest and harbor countless colonies of creatures. More than 175 species of birds and animals, 700 plants, and 3,000 insect species are known to use downed logs for food, homes, or shelter. For many plants, beginning life on top of these logs affords a better chance at survival because they boost new seedlings above the forest floor and away from competitors. For example, shade-tolerant western hemlocks commonly begin life in colonnade rows atop mossy, partially decomposed nurse logs. In a few years, they run their roots down the log's sides to the soil below. As the "mother" log deteriorates, these hemlocks mature and eventually a row of stilt-like trees remain, roots holding the trees' trunks high in the air. Where competition for sunlight and space is fierce, such adaptations prevail.

Moisture is important for all forests, but the water-saturated downed logs act as enormous tanks for water storage, an important factor in maintaining constant temperature and adequate moisture under the forest canopy. As a log slowly decomposes, nutrients are gradually released into the surrounding forest. This is recycling at its natural best, using former generations of forest to build future trees, all the while creating homes, shelter, and food for current residents. These structural details are missing in managed forests because slash and debris, along with old logs and snags, are routinely burned after logging.

Downed logs in streams provide additional features. By partially blocking stream flows, logs slow water currents and create holding pools and hiding places for fish and other aquatic life. A log's nutrients leach into the stream, and woody debris becomes critical food for insects and invertebrates which, in turn, become food for fish. Downed logs are the basis of a healthy stream's structure and food chain, in that they dictate how water will flow and what will live in it. Large logs in streams account for as much as 85 percent of the organic material found in the streams. These huge obstacles form semipermanent barriers that catch other woody debris and hold it long enough for most of it to be processed by insects and bacteria. Given a choice of pools, fish always take the one with more woody debris, the one with the big downed log.

Logs in streams create pools for fish

BELOW GROUND—THE BEST IS HIDDEN

Ancient forest soils are teeming with life, having developed undisturbed for thousands of years. Here are complicated interactions of fungi and tree roots. Hyphae—the living, growing bodies of fungi—are everywhere, interacting with tree roots and other plants to form a common bond. This association is called *the mycorrhiza process,* and it links hyphae with plant roots for a two-way exchange of nutrients and water. A cubic eighth-inch of soil or rotting wood can have more than 300 linear feet of these tiny fungal tubes. A square meter can have 2,000 earthworms, 40,000 insects, 120,000 mites, 120 million nematodes (such as pinworms or hookworms), and an almost uncountable number of bacteria and protozoa—all interacting, eating, releasing wastes, and reproducing. This is a busy place, busier than most people realize because the majority of these processes are hidden from the human eye.

Conifers

Nowhere else on earth are there trees like this. There may be older trees or larger trees in other forests, but as a group, Northwest forests contain trees unsurpassed in both size and age. Of the eleven dominant conifers occurring in the Northwest, six commonly exceed ages of five hundred years and three ages of three hundred years. Before logging claimed the best, trees larger than 10 feet in diameter, 200 feet in height and five hundred years of age were fairly common, not the rare record holders that are honored today. The Pacific Northwest provides an almost perfect home for these giants, where moisture and temperatures provide nearly a twelve-month growing season.

Conifers are trees with needle-like or scale-like leaves that typically remain on the tree throughout the year, unlike their broadleaf relatives that usually lose their leaves in winter. Rainfall in the Pacific Northwest is not only high, but also occurs mostly in the fall, winter, and spring, the opposite of most climates. This pattern handicaps broadleaf trees that rely on the warmer but drier summer months to achieve most of their photosynthesis. By contrast, conifers complete much of their photosynthesis during the wet months, when sunlight and temperatures limit growth but moisture does not. The sheer size of these trees also aids them, as they can store huge quantities of water for use during the dry summer season. In such conditions, conifers win out and broadleaf trees are relegated to second-tier status as streamside or midcanopy trees.

DOUGLAS-FIR (*Pseudotsuga menziesii*)

Scottish physician and naturalist Archibald Menzies first described Douglas-fir in 1791. Around 1827 botanist-explorer David Douglas also found this tree and brought seeds back to England. Botanists first mistakenly classified it as a pine, but it doesn't have needles or cones. The pointed needles look somewhat like spruce, and the bark looks like that of a fir. Thus the tree was called Douglas-spruce for awhile; it was also referred to as Oregon-pine and classified as a true fir and a hemlock. Confusing, isn't it? Today it is simply known as Douglas-fir. It is classified as *Pseudotsuga* (meaning false hemlock) and *menziesii* to acknowledge its discoverer.

Arriving in the Northwest about seven thousand years ago, after climates warmed and ice from the last ice age melted, Douglas-fir has since dominated most forests where the conditions are somewhat warmer and drier than on the coast. In the Northwest's moist climate Douglas-fir grows rapidly, reaching heights of 200 feet or more and diameters of 4 to 8 feet. The corky bark may be as much as a foot thick, providing admirable insulation against wildfire. Commonly used as Christmas trees, the trees, with their yellowish-green 1¼-inch flat needles, have a distinctive resinous fragrance. An old-growth Douglas-fir may have 60 million needles representing a surface area fifteen times the ground area beneath it.

Douglas-fir cones are unique: the small, 3-inch hanging cones have protruding, paper-thin scales. Beneath each of these, an odd three-pointed bract hangs, looking much as if a small mouse had crawled inside and left its two hind feet and tail still exposed. The winged seeds are small, with 43,000 seeds per pound, and studies have shown that they can be spread as far as a quarter-mile away by a gentle breeze. Douglas squirrels, chipmunks, mice, and shrews, as well as crossbills, winter wrens, and other birds dine on these seeds. Deer eat new shoots; bears often strip bark and eat the sap. Red tree voles, tiny mouselike creatures,

Douglas-fir

High overhead, ravens soar through the forest

live out generations in Douglas-fir treetops, eating needles and seldom coming to the ground. A fifth of the foliage weight of an old Douglas-fir may be lichens, which eventually provide nutrition for the forest's many creatures and other plants.

Douglas-firs are among the world's tallest trees, with the current champion in Coos County, Oregon, more than 329 feet tall . Historic records indicate these trees probably exceeded 400 feet in height in the best-watered lowland sites. The trunks on these old trees are also impressive, with diameters of 13 to 14 feet. Longevity is another trait, with many trees living 750 years, and some living to be 1,000 years. This mighty and interesting tree species, which once towered over lands now called Portland, Seattle, and Vancouver, is mostly gone.

Fire is important for the continuation of this species. Intolerant of shade, Douglas-firs are eventually succeeded by redwood, western hemlock, or red cedar, all of which tolerate deep forest shade well. Without a good fire occasionally clearing away the forest canopy, Douglas-firs eventually lose out to these other trees. Enormous old trees in a forest represent memories of fires that swept the area long before.

ALASKA CEDAR, WESTERN RED CEDAR, AND PORT ORFORD CEDAR

The Northwest's First Peoples called cedar, specifically western red cedar, the "tree of life." From cradle to coffin it was the cornerstone of their culture, and actually delineated the Northwestern peoples from others, as the easily split and yet rot-resistant cedar was used for canoes, house planks, posts, totem and mortuary poles, boxes, baskets, clothing, and hats. Cedar was also used for dishes, arrows, harpoons, masks, rattles, combs, fishing floats, whistles, paddles, bandages, and towels. It proved an excellent form of fuel and was used for heating, cooking, and smoking. The power of cedar was said to be so great that merely standing with one's back to a cedar gave one spiritual and healing powers. All cedars have scale-like, feathery hanging needles and fibrous bark that loosens in vertical strips. Because of their unique characteristics, cedars are unlikely to be confused with other tree species.

WESTERN RED CEDAR (*Thuja plicata*) prefers moist to wet, shady forests at low to middle elevations. Although it occurs in drier areas, red cedar flourishes in wet, boggy areas where wildfires are rare and groundwater is abundant. The largest trees are less than 200 feet tall, but buttressed bases make them appear enormous; they can have circumferences of more than 50 feet, rivaling those of redwoods. The current red cedar champion is only 159 feet tall but has a girth of more than 63 feet. Several red cedars in Olympic National Park have diameters of more than 20 feet. These trees may live to be a thousand years old. Seed cones are small and egg-shaped, about ½-inch long, with very small seeds; it takes 400,000 to make a pound. The tops of the trees droop somewhat (hemlocks treetops do this too), and the J-shaped branches droop and then sweep upward. Western red cedar is found from just north of Petersburg, Alaska, down to and just intermingling with the coast redwoods of Northern California.

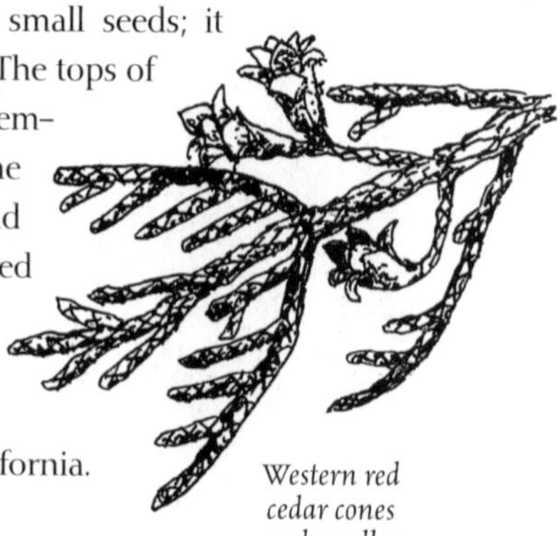

Western red cedar cones and needles

ALASKA CEDAR (*Chamaecyparis nootkaensis*) is found from Southeast Alaska to southwestern Oregon, with a few scattered groves in the mountains of California. From tideline to timberline this tree spans two thousand miles of coastline. Alaska cedar (also called yellow cedar) is the oldest tree in this area, with many living to be more than one thousand years. Being somewhat intolerant of shade, however, these cedars rarely dominate a forest; instead, they find habitats where few other trees thrive, such as those with too much moisture or rocky outcroppings. The Alaska cedar has sparse branches and a somewhat droopy form. Crushed boughs smell a bit like mildew, unlike the pleasantly scented red cedar. Another way to tell the two species apart is to stroke the foliage against the grain; Alaska cedar is prickly, while red cedar is not. Seed cones are round, bumpy, light-green juniper-like berries that ripen to brownish cones.

PORT ORFORD CEDAR (*Chamaecyparis lawsoniana*) prefers lower elevations in the coastal valleys of southern Oregon and northern California. Its pure white wood has made it highly valued, and its pleasant smell rivals that of the most expensive perfume. The current champion is in Northern California's Siskiyou National Forest and is 219 feet tall with a 37-foot circumference at 4½ feet. The branches are similar in appearance to those of the Alaska cedar, but a whitish powder appears on the tiny scales with a whitish X below each one. The genus name *Chamaecyparis* actually means "false cypress" because the cones resemble those of cypress. As if the

Alaska cedar *Port Orford cedar*

insatiable demand for Port Orford cedar weren't enough of a burden on the species, Port Orford cedar root rot has now been introduced to these forests (the fungus is spread by vehicle tires), further opening questions about the future of this magnificent tree.

COAST REDWOOD (*Sequoia sempervirens*)

Closely related to the giant sequoia (*Sequoia giganteum*), the largest living thing on the planet, the coast redwood is slimmer and taller than its Sierra kin. These two trees are relics of a time when the prevailing weather was both warmer and wetter. As climates changed, however, both redwood species reduced their range to their present California areas. A third species, the dawn redwood, is a relatively small tree and grows in the foggy valleys of China. If it weren't for the fogs that keep the coast redwoods watered and cool during California's summer droughts, these trees might not exist at all. The pre-settled forest ranged from Big Sur north to just over the Oregon border, within view of the coast (but not on the beachfront), and to about 30 miles inland. Apart from coastal fogs, redwoods prefer pocket valleys where groundwater is abundant and nighttime fog accumulates. The largest and most impressive of the remaining redwoods are within 90 miles north and south of the Eu-

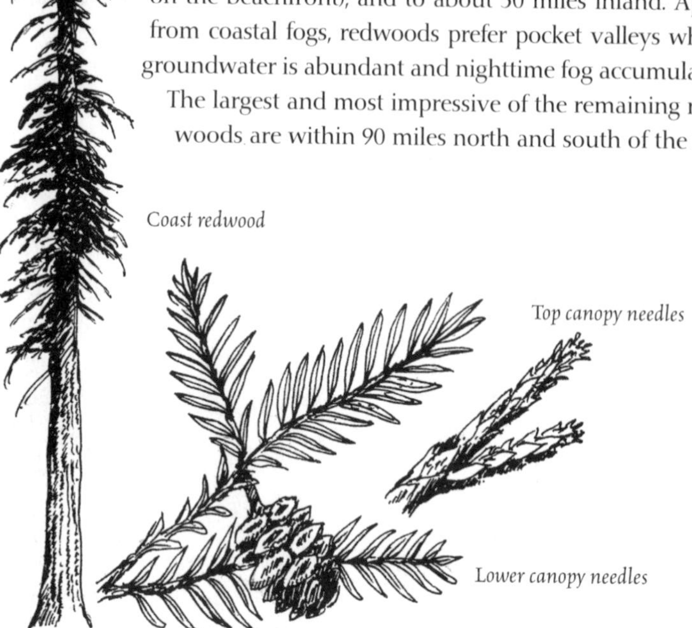

Coast redwood

reka area. Of the 2 million acres of redwoods growing 150 years ago, less than 4 percent remain today. Most old-growth trees are now protected, but trees are still being cut and protesters are arrested and sometimes even killed. The Save-the-Redwoods League is primarily responsible for the redwoods still standing today, having purchased more than $5 billion worth of redwood lands since its founding in 1918. This is a tree species that stirs many people to open their awareness—and their wallets.

Towering straight and tall to more than 350 feet, yet with an extremely shallow root system only 6 to 8 feet deep, redwoods can completely dominate river flats in lowland forests. This is where the biggest trees tend to cluster. Hillside trees are smaller and share the forest with Douglas-fir, tanbark oak, and madrone. In the northern redwood regions western hemlock and Sitka spruce mingle in boggy lowland groves, showing a connection with more northerly old-growth forests.

Redwood is an incredibly beautiful and long-lasting wood, but structurally it is very soft. Tannic acid replaces sap, providing a toxin that almost eliminates insect pests. The thick bark lacks true sap, and the clear trunks that rise many hundreds of feet without branches or foliage help the tree to resist fire and assure that redwoods live long lives. The oldest recorded redwood was twenty-two hundred years old when cut. This remarkable duration continues long after a tree falls, however, for many downed logs last centuries longer. Redwoods so dominate their surroundings that many groves appear open and cathedral-like, more spacious than other complex old-growth forests that have a variety of trees.

The redwood cones—about the size of an olive—seem absurdly tiny for a tree of this stature. The seeds look like those of a tomato, with 132,000 seeds per pound. The needles are unique, with upper-canopy foliage looking more like spikes of juniper while lower-canopy boughs are flat and broad. This is a result of the different levels of moisture loss and available sunlight in the two areas that are hundreds of feet apart. A single redwood can "exhale" more than 500 gallons of water a day. Many redwoods have been proclaimed the "tallest" and the "champion" in the past few years because accurate measurements are difficult to obtain for trees this tall. The tallest one recorded today may be surpassed tomorrow. Suffice it to say, this species is the tallest tree in the world, with many individual trees more than 350 feet tall.

SITKA SPRUCE *(Picea sitchensis)*

The fourth-tallest Northwest tree species, this "Tideland spruce" favors the Pacific shoreline from Southeast Alaska to the coast redwoods of Northern California. It is rarely found outside the fogbelt along this two thousand miles of coast, where climates are cool and wet. In northern California, Sitka spruce abruptly stops north of foggy Cape Mendocino, with a small remnant grove at Point Arena, just north of San Francisco, a relic of wetter times.

Projecting needles on all sides of spruce twigs are sharp and three-sided, flat on top. Buttressed bases of trunks add support, and flexible twigs yield to coastal winds. The finest boat masts and airplane parts were once made of Sitka spruce, owing to the flexible qualities of this coastal tree. Purplish or reddish-brown bark flakes off in scales, unlike any other Northwest tree species. Downward-hanging papery-scaled cones are about 2½ to 4 inches long, and there are 210,000 tiny seeds per pound. The largest Sitka spruces commonly tower higher than 300 feet and have trunks that may be more than 17 feet in diameter, but there are few of these left. Fast growing, they can add a foot of girth in a decade. These trees handle shade well, and many get their start on forest nurse logs. They may live to be older than 750 years. Market demand for tight-grained old-growth Sitka spruce wood has destroyed most of the finest trees. It is rare to find an ancient spruce still standing.

Sitka spruce

Western hemlock

WESTERN HEMLOCK (*Tsuga heterophylla*)

Probably the most abundant conifer from coastal Oregon to Southeast Alaska, western hemlock is the largest of the world's hemlocks. Soft and delicate seedlings often blanket nurse logs and forest floors. Being very shade tolerant, hemlocks can persist for long periods waiting for dominant trees to die. Hemlock's droopy treetop and the small, soft, downward-hanging branches and cones allow easy identification.

Three champion trees are located in Olympic National Park, with the largest reaching 241 feet high. Although hemlocks are tall, few have enormous diameters like their neighboring conifers. Slow growing in shade, hemlocks eventually win out over other tree species to become the dominant conifer of most northern coastal old-growth forests. They are prolific, but because they lack thick bark and strong roots, hemlocks are vulnerable to fire and windstorm. Their elliptical cones are about an inch long and hold paired winged seeds that aid in dispersal under a windless forest canopy. A full-grown hemlock drops more than one seed per square inch of ground beneath it. Of course, only a small fraction of these ever mature into adult trees.

TRUE FIRS

The true firs of the Pacific Northwest (unlike the Douglas-fir) are often difficult to distinguish from one another. Extensive intergrading and variable needles make these three (the grand fir, the Pacific silver fir, and the noble fir) very confusing to identify.

GRAND FIR (*Abies grandis*) is scattered in forestlands considered slightly drier and below 2,000 feet in elevation. It prefers inland valleys and rain-shadow areas between British Columbia and Northern California. Also called lowland white fir, this tree may reach 200 feet high with a trunk 4 feet in diameter. The glossy green needles are broad and thin, spreading on a flat plane from the twig. They are dark green above with two white stripes on the underside. The species is most common near streams and is often found with cottonwoods and red cedars. In Northern California it may hybridize with its close relative, the white fir (*Abies concolor*).

PACIFIC SILVER FIR (*Abies amabilis*) is also called the Amabilis fir or lovely fir. A striking tree of midelevation snow zones, from 3,000 feet to 5,000 feet in Oregon and down to sea level in southeastern Alaska, the largest tree in this species is more than 200 feet tall and almost 8 feet in diameter (found in Olympic National Park at the relatively low elevation of only 800 feet). It is a very shade-tolerant tree, as shown by the flat, spreading needles that are designed to catch as much sunlight as possible. Cones sometimes stand upright on upper branches sometimes for several years, dense and barrel-shaped, 3 to 5 inches long; they expel their seeds from the top first, creating a sort of candelabra effect.

NOBLE FIR (*Abies procera*) is closely related to the California red fir (*Abies magnifica*) and intergrades into the Shasta red fir (*Abies magnifica shastensis*) of Northern California. This "noble" tree occupies a relatively small range when compared with most other lowland old-growth trees. It is the largest and longest-living true fir in the Northwest, growing taller than 200 feet, with a diameter of more than 4 feet. While noble and Pacific silver firs don't grow in what are commonly termed *lowland old-growth forest*, they add to the overall composition of ancient forests of the Pacific Northwest.

Grand fir
Pacific silver fir
Noble fir
Typical shape for all three trees

PACIFIC YEW (*Taxis brevifolia*)

The Pacific yew is an odd tree; it is a conifer but it doesn't have the abundant needles of most "evergreen" trees. It doesn't have cones; rather, it bears seeds singly in berry-like arils—and only on "female" trees. These attractive fruits are toxic enough to produce cardiac arrest in humans, but birds enjoy the berries and are unharmed by them. Most people would classify conifers as "softwood," but the rose-orange heartwood and cream sapwood of the Pacific yew is among the hardest of woods.

Native Americans used yew for countless utensils—from spoons to clam shovels, from drum frames to bows and arrows. Timber folks called it a "trash tree" and regularly burned it with other "slash," thinking it was useless. In 1987, however, it was discovered that taxol, a compound extracted from yew bark, shrinks a variety of cancer tumors, and the yew's reputation changed. Suddenly, this lowly secondary-tier tree—the only Northwest conifer that doesn't achieve upper-canopy status—became famous. Once bulldozed down as unmarketable, passed over and burned as a weed, the Pacific yew became the timber target of the 1990s. Things didn't look good for this odd and unique tree, but thanks to science, taxol has now been synthesized, and the rush to harvest yew bark has eased. This is a classic example of the secrets these forests hold; undoubtedly there are more.

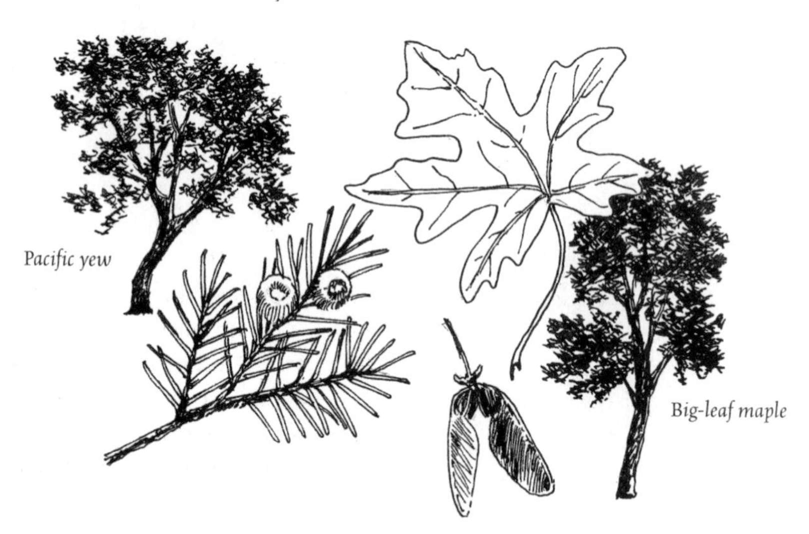
Pacific yew

Big-leaf maple

Broadleaf Trees

Forests of the Pacific Northwest are unique in their overwhelming dominance of conifers relative to other tree species. Although the Northwest's giant conifers capture the lion's share of fame for their astounding size, broadleafed or hardwood trees (those that generally loose their leaves in winter) are also impressive in their own right. But conifers overwhelm hardwoods by a ratio of about one thousand to one. Usually thought of as an "understory" or a marginal forest component, they are enormous when compared with other broadleafed trees in the eastern United States.

BIG-LEAF MAPLE *(Acer macrophyllum)*

Giant leaves 8 to 12 inches across with five deep lobes distinguish this interesting and beautiful Northwestern tree. Short trunks divide, reaching 70 to 80 feet tall; with wide, open crowns that are just as wide. Big-leaf maples often live two hundred years. Winged seeds, as well as twigs, buds, and leaves all grow in pairs opposite each other. The most famous big-leaf maples are undoubtedly those found festooned with mosses in the westside forests of Olympic National Park. Each maple is lavishly draped with more than a ton of moss and ferns; these trees epitomize the West Coast rain forest. It has recently been discovered that moss, ferns, and old leaves build up a soil of sorts on the maple's broad branches. Concealed beneath the mosses, tiny tree roots grow from branches into these nutrient-rich moss gardens, gleaning nourishment the mosses and ferns have created. The mosses and ferns in turn gain from this by enjoying less competition for space than the earthbound plants below.

BLACK COTTONWOOD *(Populus trichocarpa)*

Easily the tallest and most massive broadleaf tree of the Pacific Northwest, black cottonwood prefers moisture-rich streamsides and riverbanks. Towering to 175 feet tall, with trunks 4 to 5 feet in diameter, these cottonwoods have interior wood that seems almost spongelike, and it can be so wet that water actually spurts out of a cut stump. The broad triangular leaves and thick, deeply furrowed ashy-gray bark distinguish

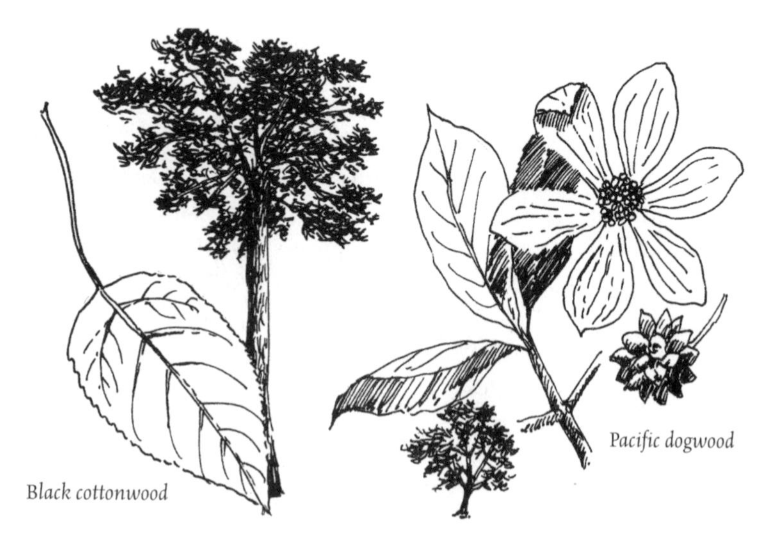

Black cottonwood

Pacific dogwood

black cottonwood from any other northwestern tree. Rotten cottonwood trunks are important to wildlife, as they provide homes for a great variety of birds and animals. The largest known cottonwood is located near Salem, Oregon, and has a trunk 9½ feet in diameter.

PACIFIC DOGWOOD *(Cornua nuttallii)*

Undoubtedly the most beautiful flowering tree of old-growth forests, dogwood "flowers" are not what they seem. The true flowers are small, greenish, and rather drab. The surrounding clusters of bracts, however, are creamy-white and create the radiant blossoms mistakenly called flowers. By fall the flowers have ripened into bright red fruits about one inch across, showy indeed when the leaves turn orange, red, and purple. Dogwood is unusual in that it is commonly found beneath all of the giant conifers, from coast redwood to Sitka spruce and Douglas-fir forests. This is a shady habitat in which few other trees are able to survive.

Other Plants

Flowers, ferns, fungi, and other plants may seem to be less compelling components of the old growth forest when compared with the tree giants, but they are just as vital to the overall ecosystem. In fact, without the lowly fungi, trees might not even be able to grow. Many thousands of species of plants and creatures use old-growth

A riot of spring flowers dwarf a Pacific treefrog. Such a scene would not exist in fiber farms

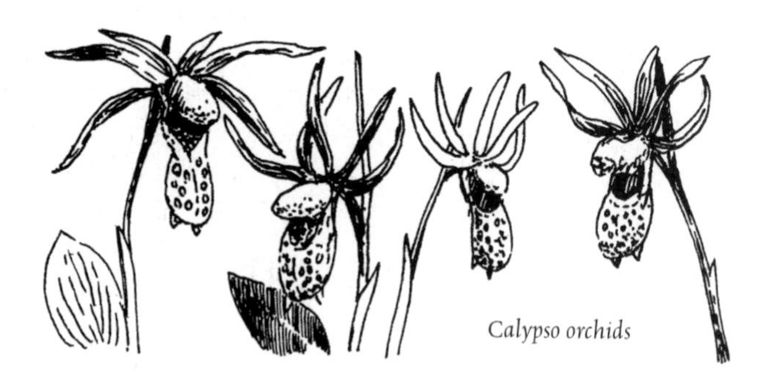
Calypso orchids

forests; some of these have important connections to the forest or are so unique in their requirements that they can live nowhere else. Some species require the heat-regulating temperate qualities of dense forests, such as spotted owls or delicate forest flowers. Others, such as salamanders and frogs, require a constantly damp environment. As these types of older forests continue to fragment and disappear, the future of many of these plants and creatures that are dependent on the qualities of old-growth forests grows increasingly uncertain.

LOWER CANOPY SHRUBS: HALFWAY BETWEEN

Beneath the forest canopy, where billions of needles hog most of the sunlight and where the ground is colonized by a drapery of moss and flowers, many graceful and meager shrubs reach for the little remaining sunlight. These are mostly broadleaf shrubs that have adapted to a half-light landscape, where rain and snow are filtered and wind rarely blows.

DEVIL'S CLUB (*Oplopanax horridus*) rises out of shady, boggy, wet thickets. Ten-inch red-fruited spikes ascend above huge leaves; they are beautiful except for hidden spines that accompany the plant's stout stems. Thorns protrude from the undersides of the leaves' main stems, making this plant's name more than appropriate. Related to ginseng, devil's club was and still is used by Native Americans as an important medicinal agent, to treat such ailments as arthritis, colds, and ulcers. **VINE MAPLE** (*Acer circinatum*) is a small, vinelike cousin of big-leaf maple, known to many hikers for its vivid and incandescent fall color in many lowland forests. **BLACK** and **RED HUCKLEBERRY** (*Vaccinium* ssp.), **SALMONBERRY** (*Rubus spectabilis*), and **THIMBLEBERRY** (*Rubus parviflorus*) are common bushy and erect shrubs in old-growth areas. They all fare better in increased sunlight; some approach weedy status in cutover landscapes. Berry fanatics debate the palatability of all of these fruit-bearing shrubs, but the thrushes prefer the salmonberries.

FLOWERS OF THE ANCIENT FOREST

Bewildering arrays of flowers bloom during the prolonged springtime of old-growth forests. So moderate and even are temperatures under the canopy that in many places spring begins in February and continues through July. Dainty and delicate, fragile and frail seem appropriate descriptions for most of these plants that depend on shady and moist conditions. This is not the sort of place one will likely see such sun-loving flowers as thistles or dandelions. Non-native species such as these have crowded out many native flowers in other habitats, but here in the old-growth forest of the Northwest, it is habitat destruction that has made some native plants exceedingly rare.

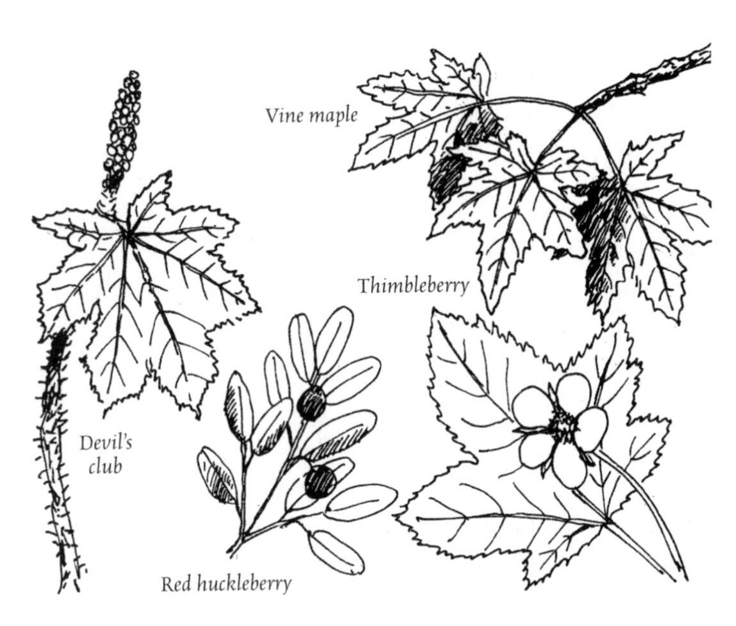

Covering the floor of many redwood and coastal forests, **REDWOOD SORREL** (*Oxalis oregana*) looks something like common three-sided shamrock or cloverleaf. Horizontally flattened leaves catch all of the available secondary light, but they quickly fold down vertically to prevent dehydration when direct sunlight hits them. **INSIDE-OUT-FLOWER** (*Vancouveria hexandra*) is a delicately stemmed plant with small white flowers that appear to turn themselves almost completely inside out. In deep forest, where plants cannot depend on wind to transport seeds, wasps and ants sometimes disperse these small fleshy grains. **BUNCHBERRY** (*Cornus canadensis*) hugs the forest floor with oval leaves in whorls of six. The white outside bracts are often mistaken for flower petals, but the true blooms in the centers are tiny and inconspicuous, turning to red-orange berries in fall. The flowers look just like tiny dogwoods for good reason: Dogwood is bunchberry's closest relative.

TRILLIUM, ANDREW'S CLINTONIA, FAIRY-BELL, FALSE LILY-OF-THE-VALLEY, BEADLILY, SAXIFRAGE, CANDYFLOWER, WILD GINGER, SOLOMON'S SEAL, QUEEN'S CUP, STARFLOWER, COLUMBINE, IRIS, TIGER LILY, CALYPSO ORCHID, TWINFLOWER, and **BLEEDING HEART** are just a few of the many wildflowers found in old-growth forests. All of these require shady landscapes; almost all disappear when forest disappears.

Redwood sorrel

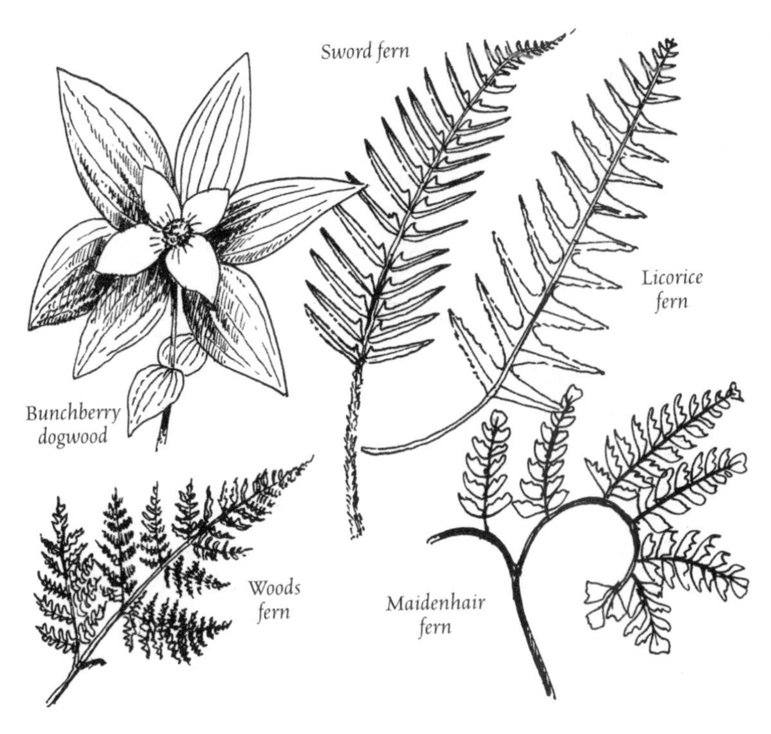

A FOREST OF FERNS

During the Mesozoic period, 230 to 70 million years ago, fern "forests" covered much of the planet, and many ferns were larger than current sizes. Today, ferns are subordinate players to old-growth trees but countless numbers grow in these regions, creating impressive groundcover—a "canopy" just above the mosses and flowers. In some forests ferns are the most common and dominant understory plants. Sword, lady, and woods ferns and bracken, deer, maidenhair, and licorice ferns are a few of the more common species found in old-growth forests. The evergreen sword fern is one of the most abundant and cosmopolitan plants of the ancient forests, and can readily be identified by the little sword "hilt" on each pinna section of frond, some plants may reach 5 feet tall. Licorice fern (so named because its rhizomes have a strong taste of licorice) has taken a different approach, commonly growing on maple or alder logs or tree trunks and preferably in a bed of mosses to keep its roots wet.

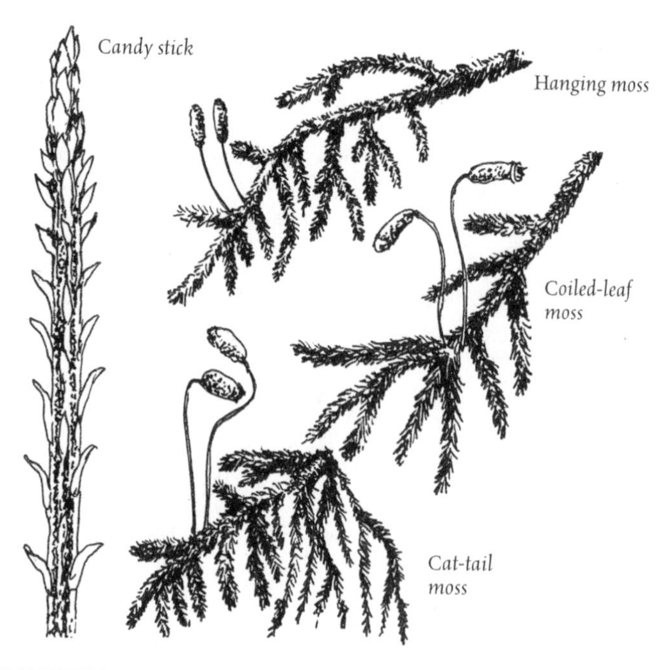

HERBS WITHOUT CHLOROPHYLL

An odd group of old-growth plants, the epiparasites (misnamed "saprophytes" by some) do not make their own chlorophyll. Defining these oddities simply as spikelike flower stalks with shrunken or nonexistent leaves doesn't really explain the strange connection these plants have with old-growth forests. Take the candy stick (*Allotropa virgata*) as an example: The entire foot-high stalk is astonishingly bright red and white striped. Beneath the ground these plants obtain their nutrients from living fungi that are hooked up with plant roots, usually those of trees. The exchange of nutrients at root level, through organs called "mycorrhizae," allows fungi and plants to take advantage of each other's strengths. The candy stick just taps into this connection, however, and doesn't appear to give anything back in return. It is a flower stalk, using the tree's leaves for photosynthesis and the fungi for its roots. In an otherwise intertwined forest ecosystem, this is difficult to comprehend. Pinedrops, pinesap, Indian pipe, and ground cone are some of the other epiparasites in these forests.

MOSS, LICHENS, AND LIVERWORTS

The famed Hoh Rain Forest in Olympic National Park is probably better known for its moss-festooned trees than for the trees themselves. Hanging moss gardens of green drapery descend almost to the ground and give a carpeted and curtained look to the place. These mosses, liverworts, and lichens are called epiphytes (plants that grow on other plants), and are some of the most unique and interesting ingredients of old-growth forests. Mosses and liverworts are nonvascular plants, that is, they have less-developed water- and food-conducting systems than do flowers, trees, or ferns. The immense number of species of epiphytic plants found in the Pacific Northwest indicates that this is an ideal climate for them.

There are about seven hundred species of **MOSS** in the Northwest, and to many people these species all look the same. There are many differences, however. Some mosses colonize specific surfaces such as rocks; some prefer hardwoods; others like downed logs, disturbed or undisturbed soils, burned tree stumps, or upturned tree roots; still others flourish out of water, beside it, or directly in it. There seems to be a moss for every occasion.

Wet old-growth forested areas tend to have **HANGING MOSS** (*Antitrichia curtipendula*), one of the characteristic moss "balloons" that decorate tree trunks and branches. Rusty to orange-green in color, the large, loose cushions and mats seem to cover every inch of available surface. Two other common rainforest mosses are **CAT-TAIL MOSS** (*Isothecium myosuroides*) and **COILED-LEAF MOSS** (*Hypnum circinale*). Rather confusing to identify specifically, both species are glossy green, but cat-tail moss has straight, sharply pointed leaves and coiled-leaf moss has leaves that curve strongly to one side. Confused? With upward of two thousand species of mosses, liverworts, and lichens in the Northwest, even the experts have a difficult time of it. **GOOSE-NECKED, PIPE-CLEANER, BENT-NOSED, OREGON BEAKED, SPEAR, GOBLIN'S GOLD, LOVER'S, AND GLOW MOSSES** are a few common names of Pacific Northwest mosses.

LICHENS are actually composite organisms composed of simple algae and fungi living together. More than one thousand species make their home in the Pacific Northwest. Think of lichens as fungi that are farmers. Instead of scavenging a living for themselves the way such fungi as molds, mushrooms, or mildew do, lichens cultivate algae within themselves. Algae are photosynthesizers that provide the fungi with carbohydrates, vitamins, and proteins. In return, the fungi give the algae a safe place to live. This mutual relationship is the classic symbiotic association, with each plant providing what the other cannot: Together they live, divided they fail.

LUNGWORT (*Lobaria pulmonaria*) and **LETTUCE LUNG** (*L. oregana*) are both large-lobed and loosely attached foliose (leaflike) lichens associated with old-growth forests. Commonly found growing in treetops and often hidden by lower branches, they may approach quantities of 500 pounds (dry weight) per acre. These species may account for about a fifth of the total foliage biomass of an entire old-growth Douglas-fir. These lichens grow slowly, and in one hundred years may reach the size of a small piece of lettuce. As they grow, they "fix" atmospheric nitrogen from air; because few plants can obtain nitrogen this way, these lichens are important sources of this essential chemical. Rain leaches nitrogen from lichen, drenching trees, plants and soil below with a chemical soup. Winter storms also blow bits of lichen to the ground, where eventually it is either eaten or decays, adding important fertilizer to the forest. Few lichens are found in young forests.

Foliose lichens

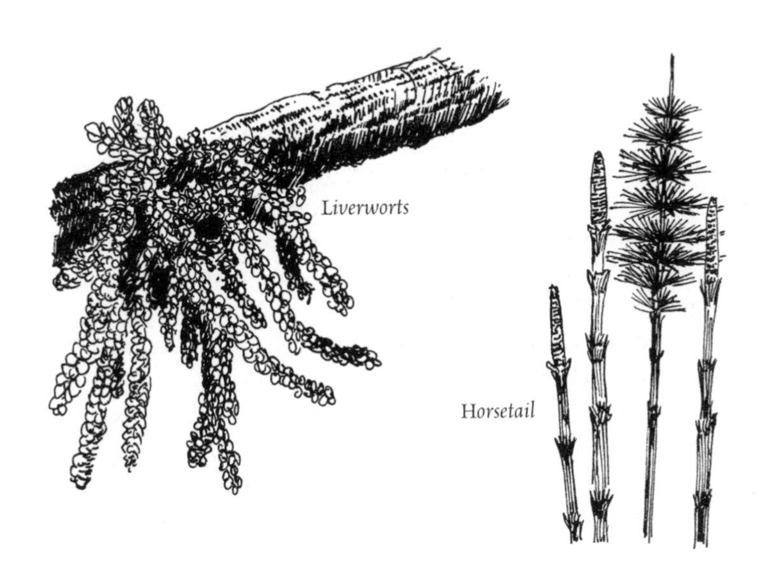

FUNGI

Mushrooms are the fruiting bodies of fungi, much like apples are for an apple tree. Mushrooms are small and temporary parts of the entire below-ground plant that is really an enormous interconnected network of fine hairs, the mycelia. Fungi lack chlorophyll and therefore must rely on organic material for their nutrition. Mushrooms produce spores that spread the fungus to new locations, much like a flower produces seeds.

Fungi are everywhere in the old-growth forest, from tree roots to the treetops, and they provide connections with other plants and creatures. In the canopy fungi join algae, yeasts, and bacteria to form a microscopic salad on and within old evergreen needles that forest scientists simply call *scuzz*. Some of these fungi are toxic to needle-eating insects (thereby protecting the host tree), while others capture nitrogen from the air and eventually pass it along to the trees. In forest soils fungi form interconnected or mycorrhizal relationships with tree roots, providing moisture and nutrients to the trees, while the trees pass sugars and amino acids back to the fungi. In this manner trees and fungi are intertwined. It appears that all Northwest conifers have mycorrhizal fungi connections.

Some fungi may be as endangered as salmon

Like salmon in the Northwest, fungi may also be endangered; some Northwest fungi species may already be extinct. Not enough is known about them. As an example, an extremely rare fungus known as "fuzzy sandoni," or *Bridgeoporus nobilissimus*, has only been found six times, all in old-growth noble fir, silver fir, and western hemlock. *Nobilissimus* means "most noble" and describes what appears to be the largest fungus in North America, possibly the biggest in the world. One such fungal fruiting body was 55 inches across and weighed 300 pounds. For many years it was registered in the *Guinness Book of World Records* and was the first fungus to be listed as endangered. No one knows if this fungi species has gone the way of 90 percent of the region's old growth, or if it still exists.

Critters of the Forest

Some Pacific Northwest forests support more than eighty animal species, including such distinctive ones as the **RED TREE VOLE**, the world's only exclusively needle-feeding animal. A closely related **WHITE-FOOTED VOLE**, also found nowhere else on earth, is here as well. The **NORTHERN FLYING SQUIRREL** (the world's only lichen-consuming rodent) and the tiny **SHREW-MOLE**, the smallest mole in North America, are also only found here. Many **ROOSEVELT ELK** and **BLACK-TAILED DEER** use old growth for thermal cover and protection when the snow falls. The **AMERICAN MARTEN**, a rather stocky tree-dwelling weasel, preys on the **DOUGLAS SQUIRREL** (or **CHICKAREE**), the most numerous squirrel found in Douglas-fir old growth. Both species are absent in clearcuts. In typical "packrat" behavior, **DUSKY-FOOTED WOOD RATS** build huge piles of sticks for homes. These rats are favored food for spotted owls: In the old-growth forests, victim and prey are interconnected in a forest web of life. Several species are specifically adapted to succeed better in old-growth forests.

Northern flying squirrel

NORTHERN FLYING SQUIRREL *(Glaucomys sabrinus)*

These strictly nocturnal animals are unique in that they are the only small mammals that feed on epiphytic lichens in winter. They can't really fly, of course, but with help from skin flaps between fore and hind legs, they glide large distances, maneuver around branches, and land gently. Like voles and other animals, flying squirrels also dig out and consume fungi in spring, summer, and fall. They usually nest in old woodpecker holes and consequently reproduce more successfully in old-growth stands.

Red tree vole

RED TREE VOLE *(Arborimus longicaudus)*

The Latin name means "long-tailed tree mouse," an appropriate label for an animal that lives in the upper canopy. Few creatures are able to digest conifer needles, but this one cuts and collects branches, then consumes as many as one hundred needles per hour. It only eats the outer needle margins; the resulting debris is used to line treetop nests, which are enlarged over generations to include multiple rooms and escape tunnels. Red tree voles have been known to escape predators with a single jump and free fall from the upper canopy, legs stretched wide like flying squirrels.

BANANA SLUG (*Ariolimax columbianus*)

Most slugs evolved from snails that lost their shells when they weren't necessary to conserve moisture. Therefore, slugs need a consistently damp environment, only venturing out from under leaves and mosses on wet days. Up to 6 inches in length, the banana slug eats many plants and fungi, using a tongue covered with several thousand tiny teeth. It has been said that slugs consume more vegetation in Olympic National Park than do the park's five thousand elk. Both the slugs and the elk provide "gardening" skills that keep the forest healthy.

Banana slug

Ins and Outs of Amphibians

Amphibian means "double life," an appropriate name for creatures that move between water and land, between wet seeps, streams, and big, moist, downed logs; they are never far from water. Amphibians cannot tolerate dryness because the skins of their eggs and bodies lack moisture barriers. Their eggs hatch in ponds, lakes, and streams; during this legless larval stage in water they breathe through gills like fish. At maturity, amphibians range further afield but still depend on moist environments to help regulate body temperature. Old-growth forests and clear streams are excellent places for all sorts of frogs, salamanders, newts, and others in this family. It has been estimated that 250 salamanders occur per acre in some forests.

TAILED FROGS (*Ascaphus truei*)

These frogs are an extraordinarily ancient race of amphibians that dates from the Jurassic period, 160 to 180 million years ago. They are the last of a group once more widespread worldwide and are the only North American frogs that fertilize eggs internally. Tadpoles have mouths with specialized sucking disks that allow them to cling to rocks in swift currents. Although most frogs only take weeks to metamorphose from tadpole to frog, tailed frogs need several years. Clear cold water (in abundance in old-growth forests) is mandatory habitat for tailed frogs.

SALAMANDERS

Salamanders can be the most abundant land vertebrates in old-growth forests and are major predators of small insects, spiders, and slugs. The **PACIFIC GIANT SALAMANDER** (*Dicamptodon ensatus*) is possibly the largest terrestrial salamander in the world, growing up to 12 inches long. These are also the only salamanders with a real voice, which sounds like a short yelp or rattle. They are important predators in streams and forests and can catch mice, garter snakes, and even small fish. The **COPE'S GIANT SALAMANDER** (*D. ensatus copei*), unknown until 1970, is a close relative of the Pacific giant salamander and is strikingly marble- and slate-colored. Only a few adult specimens have been seen. It appears they seldom reach adulthood; most reach sexual maturity and breeding age while still larvae. The **OLYMPIC SALAMANDER** (*Rhyacotriton olympicus*) is chocolate brown above with flecked yellow undersides. This species generally stays in or around creeks, where it requires several years to mature. The **ROUGH-SKINNED NEWT** (*Taricha granulosa*) sports brightly colored skin that produces toxic secretions. Most

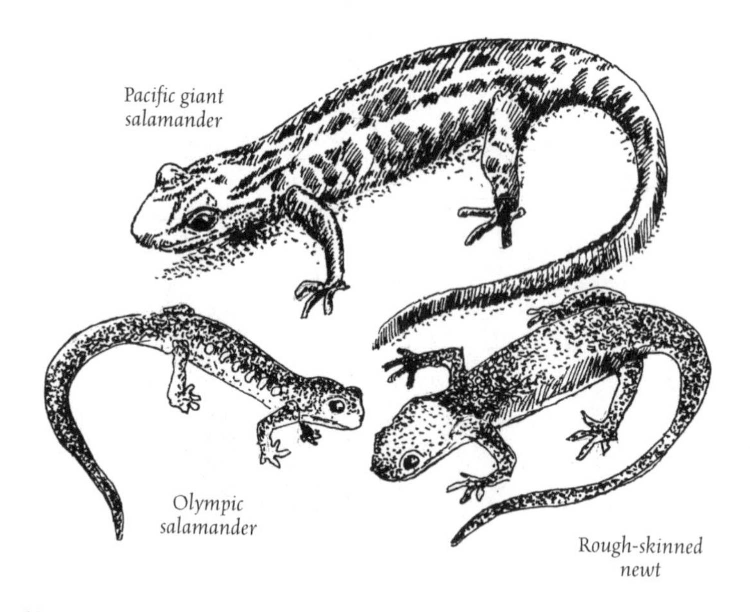

salamanders have some sort of skin toxin, but this one can sicken and even kill the creature that eats it. When threatened, the rough-skinned newts curl up and expose their bright yellow to orange undersides, a sign that reminds predators of the newt's poisonous skin.

Birds

Many bird species use old-growth snags for shelter and homes. **PILEATED** and other **WOODPECKERS** and **BROWN CREEPERS**, are just several on a long list that includes 30 to 45 percent of all Northwest forest birds. Few snags now remain outside of old-growth forests because humans have "tidied" things up. It is clear that a lack of these standing dead trees leads directly to a reduction of many forest birds.

MARBLED MURRELET *(Brachyramphus marmoratus)*

The last of North America's breeding birds to have its nesting location discovered, this robin-sized seabird remained a mystery until 1974, when a nest was found high in a Douglas-fir in the mountains behind Santa Cruz, California. Murrelets use broad branches with thick layers

Marbled murrelet

of lichens and mosses for their nest sites, and this foliage is found only in old-growth trees generally 150 years and older. Murrelets spend their days at sea, catching and transporting fish to their nests, which are sometimes 35 miles inland. In a slight mossy depression that is difficult to call a true nest, the female lays a single egg and both parents take turns incubating. They bring small fish to the nest, sometimes as often as four times a day. When the time is right, the fledgling steps off into space for its first flight, hundreds of feet above the forest floor. In 1992 the marbled murrelet was listed as a threatened species. Like salmon in the Northwest, murrelets show the clear interconnection of forest and sea.

VAUX'S SWIFTS (*Chaetura vauxi*)

These swifts rely almost exclusively on old-growth forests. Small and bullet-shaped, these acrobats spend most of their lives on the wing, chasing down and eating insects over the top of the forest canopy. They may fly as many as 600 miles a day, and some are believed to spend all of their time in the air—from fledging to building their nests two or three years later. Vaux's swifts build their nests in large, hollow, and often burned-out snags—trees often missing today from most landscapes other

Vaux's swift

than ancient forests. The swifts nest and roost in colonies, and sometimes black columns of swifts are seen ascending from a snag like a smoke cloud. Swifts have tiny feet, so small they can barely walk; they use them solely to cling to the insides of snags or other woody surfaces.

Northern spotted owl

NORTHERN SPOTTED OWL (*Strix occidentalis*)

Just twenty-five years ago, the spotted owl was one of the least studied birds in North America. Not so anymore, for after much investigation, media limelight, legal struggles, and bitter debates, many people have at least heard of this 22-ounce owl that spends its days sleeping and nights chasing rodents across the forest floor. The spotted owl was listed by the federal government as a threatened species in 1990 when it was shown that they require old-growth forests to survive. These owls nest in large live trees that have cavities or broken tops, or platforms of branches capable of holding enough material for a nest. Again, these trees are found almost exclusively in old growth. The owls prey on flying squirrels, tree voles, and woodrats—common mammals of ancient forests—and use big trees as thermal cover. Deep forest cover also makes them safer from their predators, such as great horned owls, which prefer young forests or edges of clearcuts. Studies show that spotted owls require large amounts of forest to survive—on the order of 3,800 acres of old growth in Washington state and 1,900 acres in northwestern California *per pair*! Fragmentation of forestlands is unquestionably the biggest threat to this species.

Spotted owls help the forest they live in

The spotted owl is simply one element, one intertwined component in the entire forest, no less important than any of the others. One interesting connection between this bird and several specific fungi and animal species helps to show how interconnected the old-growth web of life is. Truffles are fruiting bodies of certain fungi species, but unlike most mushrooms, truffles grow just beneath the ground's surface. Fungi play an important role in the growth of trees, providing a connection between their roots and the soil's nutrients and water. In turn, trees provide sugars and amino acids to fungi. Flying squirrels, mice, chipmunks, voles, and other creatures find truffles irresistible to eat. As these animals eat truffles, they digest the spores, which survive in their intestines and are soon deposited on the forest floor. Spores are distributed in this way and aid the growth of both new fungi and trees. Because the truffle's underground fruit cannot easily disperse spores in such normal ways as by wind, these truffle epicures may be the only way that this species can reproduce.

For spotted owls these truffle-eating diners are easy targets, and soon the owl's next meal (with its load of truffle spores) is winging its way even further afield through the forest. In this way spotted owls aid the trees by also spreading truffle spores. Remember, truffles assist the tree's roots, the very trees the owl needs for its own survival. Trees, truffles, owls, and squirrels—everything here is interconnected.

Everything works together, and smaller plants and creatures are heavily affected by changes to the forest. When logging or fire removes the upper canopy, the sun-shield of the forest floor, the impact goes far beyond simply removing the trees. Sun quickly dries the soil, creating a virtual desert. Almost immediately, sun-loving plants crowd in and replace the flowers, shrubs, fungi, and shade-tolerant plants that once grew there. Fireweed and foxglove, scotch broom and thistle crowd in, many of which are noxious aliens from overseas. Young trees soon crowd out the interlopers and, given enough time, the canopy closes overhead and native woodland species return again. This routine applies to animals and birds as well. Spotted owls are dependant on old growth, while great horned owls prefer forest edges and openings. Forest birds such as Hammond's flycatchers and winter wrens will be replaced by sunny songsters, dusky flycatchers, and house wrens.

Salmon: Fish of the Forest

No creature is more closely linked to the people and forests of the Pacific Northwest than salmon. No other creature has more human value as food, cultural icon, economic influence, and political vortex. No environmental issue has been more hotly debated, and the debate will certainly continue in the near future. Coastal runs of salmon, it turns out, are almost completely connected to old-growth forests.

Salmon, in one form or another, have been in the Pacific Northwest for a million years or so. Related to trout, salmon were originally freshwater fish, but it is easy to see how and why their relatives became migratory. Rivers and lakes in the west are relatively nutrient-poor places. Ice-age glaciers further scoured many areas, and as recently as nine thousand years ago much of this landscape was still under a vertical mile of ice—not good fish habitat! When the ice finally melted, emerging streams were nothing but gravel beds and willow bars. With so little to eat in their "home" waters, young salmon were forced to migrate downstream and were greeted by a vast and nutritious ocean. Habits of the salmon still remain somewhat of a mystery today. After spending their adult lives in the Pacific and journeying thousands of miles, they somehow manage to find their way home to the same stream they were born in to spawn and create the next generation. Seven species of salmon live in the Northwest, and many have numerous "runs," each genetically unique from the others. Most salmon spawn once and then die. They live on in various ways, helping the next generation and even the forest, creating a web of life so complex researchers are just beginning to understand it.

Salmon require a critically narrow water temperature range of between 42 and 77 degrees Fahrenheit. They are very sensitive to changes. An acute sense of smell allows them to detect probably one part per trillion or the equivalent of one drop per 500,000 barrels of water. These abilities undoubtedly aid salmon in finding the same stream of the very river they were born in. It's a navigational miracle that they find their way home.

Migrating salmon now run a gauntlet of enemies, however, and the absence of old-growth forests is a major problem. At one time most streams and lakes of the Northwest were lined with tall trees, shielding and cooling the water from summer sun. Old trees fell across the current and provided upstream pools that regulated water flows. It took years for the logs to decay, and while decaying they provided nutrients for algae, aquatic invertebrates, and eventually food for fish. As streamflows were regulated by logs, gravel settled out and created perfect spawning beds for salmon. Migrating fish also needed these holding pools to rest in and hide from predators. In summer, deep holes remained cool with high oxygen content; in winter, logs provided shelter from floods.

So what happens when the forest is removed from a salmon stream? With trees gone and shade eliminated, the water temperature rises. Without roots to hold it, topsoil slides into the stream, depositing sediments that fill holes in streambed gravel, suffocating newborn salmon, or eliminating spawning grounds altogether. When huge amounts of woody debris left from logging or development decomposes in the water, oxygen available for aquatic creatures decreases. Fish die and their food dies. With 5 to 10 percent of forests remaining, is there any wonder why salmon are in trouble?

As forests contribute to the health of salmon, so salmon contribute to the health of forests. Spawned-out fish become food for wildlife and

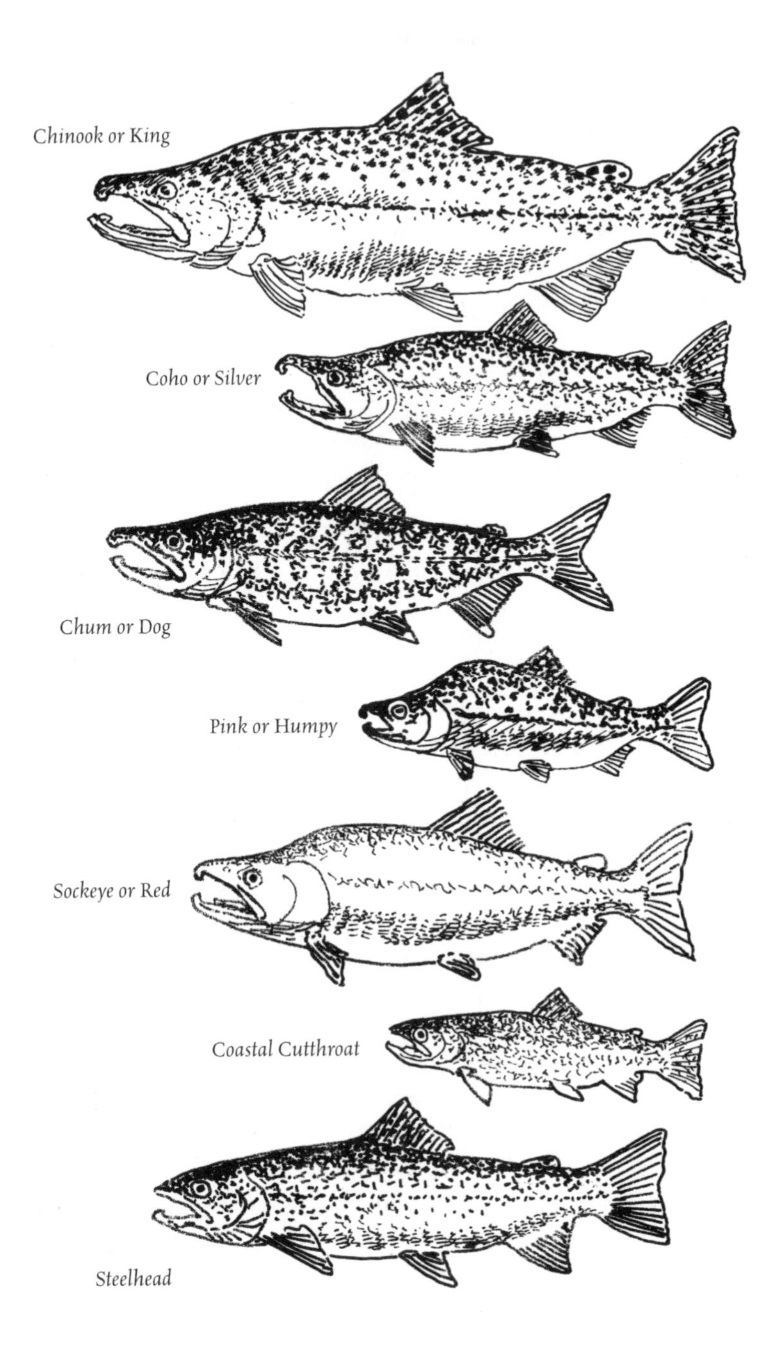
Chinook or King
Coho or Silver
Chum or Dog
Pink or Humpy
Sockeye or Red
Coastal Cutthroat
Steelhead

many arrive to feast. Bears, eagles, river otters, raccoons, coyotes, minks, skunks, bobcats, squirrels, deer mice, hawks, ravens, and crows are just a few of the diners. Many fish carcasses are dragged up into the forest, adding a rich fish emulsion to the soil and eventually to the entire ecosystem. Fish that are left to decompose in streams add important nutrients that are food for young salmon and other aquatic creatures. In this way salmon bring the ocean's bounty with them when they return to spawn, and they leave it with the forest. In death, this final gift of life benefits many. Salmon are completely and utterly dependent on a healthy forest being there when they return to spawn.

Most salmon have at least two names. **CHINOOK OR KING SALMON** (*Oncorhynchus tshawytscha*) spawn in the main stem of larger rivers, spend at least one year in fresh water and two to four years or more in salt water. They are huge fish, growing to 58 inches in length. Alaska's state Chinook record is 97 pounds, with an unofficial catch of 135 pounds. **COHO OR SILVER SALMON** (*O. kisutch*) prefer tributaries and smaller streams, spend one to two years in fresh water and one to two years at sea. Coho can jump 6 feet, allowing them to reach spawning beds above short waterfalls. Coho grow to 38 inches in length and average 6 to 12 pounds. **CHUM OR DOG SALMON** (*O. keta*) have the widest range of Pacific salmon, from Japan to Siberia to Monterey, California. From small streams they migrate downstream immediately after emergence and spend two to three years at sea. Chum grow to 40 inches in length and average 9 pounds. **PINK OR HUMPY SALMON** (*O. gorbuscha*) are the most abundant West Coast salmon. Strict in their life cycle, pinks quickly move to sea and return the next year, spawning closer to the ocean than any other salmon. They are the least dependent on fresh water and consequently are the most abundant variety. Pinks grow to 30 inches in length and 3 to 5 pounds.

SOCKEYE OR RED SALMON (*O. nerka*) are unique in their use of lakes for spawning and rearing young, but they may also use streams like other salmon. They reach 33 inches in length and 15 pounds, but most weigh between 3 and 8 pounds. Sockeye are known for their bright green heads and flaming red bodies that can be seen during spawning season. **COASTAL CUTTHROAT** (*O. clarkii*) are named for their distinctive red slash mark under the jaw and are found in small rivers and tributaries. They may spend one to four years in fresh water and as many years at sea, They are able to survive spawning several times. Cutthroat grow to 24 inches in length and average 1 to 4 pounds. **STEELHEAD OR RAINBOW TROUT** (*O. mykiss*) lack the red slash mark of cutthroat. They spend one to four years in fresh water and as many years at sea. Because they sometimes survive spawning, steelhead may return to the Pacific Ocean several times. These fish grow to 45 inches in length and weigh 5 to 10 pounds.

A Process of Preservation

"If future generations are to remember us more with gratitude than sorrow, we must achieve more than just the miracles of technology. We must also leave them a glimpse of the world as it was created not just as it looked when we got through with it."

—Lyndon Baines Johnson

Ancestors of Native Americans migrated from Asia across the Bering-Chukchi land bridge between fifteen thousand and twenty thousand years ago and fanned out across the North American continent. They reached the southeastern United States about

Native Americans used the surrounding forests for almost every commodity

fourteen thousand years ago, the Columbia Basin of north-central Oregon and the Las Vegas, Nevada, region between fourteen thousand and eleven thousand years ago. Finally, during the sequential melting of tideland glacier ice from the Wisconsin glacial stage about eight thousand years ago, they arrived in the Pacific Northwest. When they came, salmon were just populating many streams, Douglas-fir and red cedar trees were invading many forests, and much of the landscape was still raw from being covered by ice as much as a mile thick. As shorelines stabilized, sand spits and beaches began to build. It was a time of transition for the land.

When Native Americans came, the Northwest forests were decidedly difficult to penetrate and fairly impossible to travel through. Dense thickets covered river bottoms, huge logs barred forest travel, and dangerous grizzlies were common. Voyaging was mostly by water, and villages tended to hug shorelines, where there was easy access to the sea. The Northwest Native cultural heritage centers on the ocean as well as the riches of the forest. For example, western red cedar was called "the tree of life" because of all it provided: clothes, utensils, shelter, and transportation. Countless other forest plants were used as well. Prince's pine (*Chimaphila umbellata*) was a common liniment for sore muscles. Goatsbeard (*Aruncus dioicus*) was used as a cough medicine and as a salve for swellings. Thorn's from devil's club (*Oplopana horridus*) became fishhooks and lures as well as a medicine for many ailments. Huckleberry, salal, salmonberry, and many other plants were staple foods, used fresh or dried and mashed into cakes. Almost every plant had a use and each was respected for giving of itself to service. Nothing was taken for granted, nothing wasted. Much of this knowledge has been lost, however, and when a plant's medicinal capabilities (such as the components that fight cancer in yew bark or the laxative properties of cascara bark) are discovered, today's civilization is usually only copying what was previously known and used for thousands of years.

Harmony existed between human and forestlife. Simply stated, this land provided all that Native Americans required. Because human populations were small then, centuries passed with a balance between nature and people. As forests grew, generations of trees matured and died.

Eventually, seeds were sown that would became the old-growth trees of today. To walk in an ancient forest today is to enter a continuous and uninterrupted stream of life, unaltered by the heavy hand of human history. While natural changes in old growth take place on many levels today, what is there took thousands of years to develop.

CHANGES BEGAN

When Europeans first saw these forests, they understood immediately that the old, straight, tight-grained wood was enormously valuable. They had not the foggiest idea of how to turn such trees into lumber, however. Oregon Trail homesteaders and California '49ers needed lumber for homes, new businesses, and industry, but timbering first began in this region in a very small fashion. During this period, fledgling San Francisco burned down five times (most other young western cities experi-cenced fires too), and every change required more wood. By the 1870s and 1880s lumbering in the Northwest was well under way although few tools were appropriate for the task. Nowhere else on the planet had such trees been logged, and timbering technology was unprepared for the job.

Just tipping one of these forest giants over was a challenge that took several men with axes, saws, and black powder a full week. Once on the ground and bucked into sections, logs were dragged to nearby mills over wooden roads, using large teams of oxen. Smaller trees were positioned across the path and swabbed with whale or crude oil to "grease the skids,"

Oxen hauled the first logs to mill

and bull punchers used every profane word in the book to urge the oxen forward. As can be imagined with this procedure, logging occurred within a relatively short distance of small mills, and operations remained modest. Only the biggest and best Douglas-fir, western red cedar, and Sitka spruce were targeted woods in those days, and hemlock—a tree now considered common commercial lumber—was junk wood used only for skid roads. Grand fir was considered another second-rate tree, dubbed "piss fir" for its interesting odor going through the saw.

Logging history has repeated itself across North America, with the most easily accessible forests cut first. Some of the finest timber (and the easiest to cut) grew at the tidewater forests at river mouths, the inland seas of Puget Sound, and the Inside Passage. Individuals got into the act by "hand-logging," as it was called. With screws and pumpjacks and the "agility and brains of a kangaroo," any steep slope full of trees located above water put an individual into business. Two men could make good money felling trees into the nearby water and then slowly towing their trophies to the mill. Occasionally, a tree would tear down these slopes at inconceivable speed and with tremendous crushing powder, ramming through smaller trees and sending some catapulting into the air like bowling pins. Some giant cedars or redwoods could be cut and milled on the spot, to make cedar or redwood shakes.

So how did one cut down a monster tree with hand tools? Initial cuts were made several feet off the ground to get above the trunk's flared butt. Narrow planks called springboards were driven into the trunk, so fallers could stand delicately balanced to whack away with their double-edged axes. After a triangular wedge cut was made, a "misery whip" (a double-ended two-man saw up to 14 feet long) was used to make the final cut on the other side. Sometimes the biggest trees were too wide for the saw. By hollowing out an old tree's rotten interior there might be enough room for a logger to crawl inside to work his end. There were few "old" loggers.

In small stream valleys, temporary splash dams were built upstream of the "logging show." Trees were piled in the dry waterway, and when enough water had accumulated behind the dam, it was blown up, sending a cascade of water and logs surging downstream and scouring out everything in its path. Records show at least 160 splash dam sites

Donkey engines replaced oxen and horse logging

on the Oregon coast alone. Countless creeks and rivers suffered this tragic fate many times, and once pristine salmon spawning beds and fern-sided glades were turned into ravaged muddy ditches. Loggers "mined" an irreplaceable landscape, with little thought of how it would affect the forest's future. During those early days, absolutely no thought was given to the forest or its impending destruction. It was "cut and run" to the next hill. Early photographs of timber fallers beside their victims are common, showing conquerors proudly poised beside huge half-hewn trees larger than any alive today. "They'll never cut all the old growth" was often heard, as if trees a thousand years old could simply be replaced.

Over the years logging technology advanced. Horses replaced oxen because they were faster, easier to handle, and didn't come armed with horns. Then narrow-gauge company railroads replaced horses. Several inventions revolutionized the Northwest's timber industry. In 1882, inventor John Dolbeer designed the Dolbeer Donkey, a steam-powered contraption that moved logs and itself through the woods by cable. Larger versions followed, and eventually "bull" donkeys were hauling ten to twenty logs at once at a rate of 150 feet per minute. New and unique steam engines—the Shay, Climax, and Heisler—soon competed for business. They were all somewhat similar smallwheeled, geared affairs able to climb far steeper grades than their mainline rod engine rivals could. Geared engines were slow and relatively light, weighing 10 to 50 tons. With the donkey engine, which needed no tracks at all to run on, the pace of logging quickened. Northwest lumber was eventually being exported around the world, and the forests were shrinking ever faster.

EARLY OUTRAGE

As trees fell and forests vanished, not everyone was thrilled with the sight of blackened stumps and bare hills. Some early activists began to speak out. As early as the 1890s, writer and activist John Muir wrote several popular articles championing western forests and demanding government protection. He described how a lumber company hired the entire crew of every vessel that called at port. Each man was instructed to file a 160-acre claim on forest land and then deed that land over to the company. In eloquent words, Muir wrote: "Any fool can destroy trees. They cannot run away; and if they could, they would still be destroyed, . . . chased and hunted down as long as fun or a dollar could be got out of their bark hides, branching horns, or magnificent bole backbones. . . . God has cared for these trees, saved them from drought, disease, avalanches, and a thousand straining, leveling tempests and floods; but he cannot save them from fools . . . only Uncle Sam can do that."

Others were challenging the choppers as well. On May 18, 1900, while sitting around a campfire on Sempervirens Creek, just north of Santa Cruz, California, several campers grieved that the great trees they camped beneath were slated for the ax. That evening passion held sway and the Sempervirens Club was formed (it continues today). They quickly gained support and donations, some from wealthy Californians of great influence, and set about persuading an indifferent state legislature to set the forest aside for protection. Two years later Big Basin was designated California Redwood Park, the state's first official state park. The beginning of serious old-growth preservation had arrived.

In 1908, Congressman William Kent of California bought 550 acres of prime redwood forest just north of San Francisco after learning that the valley was to be cut and then flooded by a water company. It took years of effort, but this beautiful, small forest finally became Muir Woods National Monument. Other redwood forests were also given due

Logging shows are not pretty sights

consideration as being worth more than just lumber. Lumberman James Armstrong decided he just couldn't cut his extraordinary 440-acre redwood grove up in Mendocino County, California, and so it stood tall, becoming a state reserve in 1934. Recently, it was discovered to have the world's tallest tree.

In response to Muir's articles and an increasing shift of popular opinion, in 1891 Congress passed the Forest Reserve Act, giving the president dramatic power to create "forest preserves" simply by proclamation. With a stroke of a pen, forest land could be withdrawn from public domain and closed to homesteading. It was a major step in preserving western forests. This act brought about the creation of a 2,188,000-acre Olympic Forest Reserve on Washington's Olympic Peninsula in 1897 and a Pacific Forest Reserve centered on Mount Rainier in 1899. While neither reserve kept its initial size and name, these huge areas of land have become two of the largest old-growth forest tracts remaining today. Under this authority, Presidents Benjamin Harrison and Grover Cleveland withdrew and preserved almost 40 million acres, the first areas to be designated as national forests.

PRACTICAL FORESTRY AND PRESERVATION

About this time, two prominent men came to figure greatly in the battle for old growth. After several twists and turns in his government career, Gifford Pinchot became head of the Division of Forestry in the Agriculture Department under President Cleveland. Pinchot's forestry background led him to firmly believe that timberland should be "managed" by "practical forestry." His convictions were threefold. He held that forests should have profitable production rates, with scenic value of little consequence; that a constant annual yield would create steady jobs; and that forest "improvement" was necessary. Pinchot believed forests should be managed by the Agriculture Department like any other crop, and he filled his offices with like-minded people.

Meanwhile, Steven Mather, a California businessman and dedicated conservationist, had joined the fledgling Sierra Club and was soon urged by Muir to fight logging, mining, and grazing in the Sierra Nevada. While visiting Yosemite and Sequoia National Parks in 1914, Mather saw for himself cattle and sheep grazing, poor roads and trails, and private inholdings in the most scenic areas. An irate letter followed to Franklin K. Lane, secretary of the interior, who answered Mather with one of the most fateful letters in the history of conservation. "Dear Steve," the secretary wrote, "If you don't like the way the national parks are being run, come on down to Washington and run them yourself." And so at the age of forty-seven a rather naive Mather found himself the first director of the National Park Service. For $2,750 a year he was to guide and develop the national parks, something no one knew anything about. He served as director for the remainder of his career, under three presidents and five secretaries of the interior. Under Mather the National Park system expanded and preservation ethics taken for granted today were developed. Indeed from that beginning the very idea of national parks has spread throughout the world.

After Mather died in 1930, Congress quickly designated a memorial to him. This simple bronze plaque is placed in every national park, monument, and state park Mather had a hand in creating or expanding. It states: "He laid the foundation of the National Park Service, defining and establishing the policies under which its areas shall be developed and conserved unimpaired for future generations. There will never come an end to the good that he has done."

There must have been turbulent times in the Oval Office in those days, with Pinchot arguing land "management" and Mather urging

Conservationists saw scences like this vanishing

preservation of the nation's scenic heritage. Because the Forest Service controlled most of the land, Mather's Park Service wanted to create new parks, distrust, suspicion, and a bitter rivalry began that continues today. The seriousness of this battle remains obvious when one sees National Forest lands clearcut right up to the borders of National Parks.

Tree preservation continued with several major old-growth forests set aside throughout the 1920s, but protection was precarious. The original 2 million-acre Olympic forest preserve established by Cleveland in 1897 lasted only as long as it took local bigwigs to mount a lobbying effort. In 1900, President William McKinley reduced the reserve by more than a quarter-million acres, and the next year he took nearly five hundred thousand more, readying the finest timber in the United States for the ax. An additional 170,000 acres were removed from protection in 1912. Any aerial photograph of the peninsula today shows exactly where National Park boundaries are, for most of the forests surrounding the final acreage have been cut.

After World War II, logging accelerated at a frightening pace. New technologies combined with a government committed to "getting the cut out" increased logging to completely unsustainable levels. Because private lowland timber had long since been removed, remaining old growth was now mostly in National Forests in remote locations. Out of the public's eyes, these forests were managed strictly as lumber production. Little thought was given for sustaining a forest heritage or providing recreation to the citizens who actually owned the National Forests. "Forests are our Renewable Resource" was the slogan, as if ancient trees could be instantly grown again.

The 1980s ended with stringent new regulations and the collapse of many local logging economies, mostly because the trees were simply gone. The 1973 Endangered Species Act provided a legal method to guarantee the right for a species to exist. Using both the northern spotted owl and marbled murrelet as "indicator species" whose status indicates the extent of damage to the overall health of the forest, laws finally restricted logging to somewhat sustainable levels. Whether this control will continue in the future is questionable, because policies can easily can in the winds of politics. Trees will stand only by the eternal vigilance of those people passionately concerned with preserving what little old-growth heritage remains. It is clear that each old-growth forest and grove, and sometimes even single trees, remain only because a group of concerned citizens cared enough to defend them. In some cases individuals such as Muir or Mather have had profound influences. In the recent protracted and ugly battle for Headwaters Forest in California, for example, a protestor was struck and killed by a falling tree he was attempting to save.

Can these forests sustain themselves in the future even if they are not cut? Many old-growth forest creatures cannot migrate between fragmented forest homes to breed, to keep gene diversity strong. Plants cannot mix their pollen if they are many miles away from the next forest. Each remaining old-growth community can be considered a single and fragile living entity, an island surrounded by alien species and hostile conditions. As ancient forests fragment further and commercial forests and urban areas restrict the species found there, old-growth forest ecosystems will become more vulnerable to failure. There may well come a time when the health of the remaining forest ecosystems will fail, and one by one these little pockets of remaining old-growth will disappear.

Where to Find Old-Growth Forests

A bit of knowledge improves our understanding of old-growth forests, but where are they? Within a few easy hours of driving from most Northwest cities, the peace and solitude of grand forests awaits. There, experiencing sunlight streaming through the canopy, soaking in the fragrance of cedar, and hearing sounds of wind in the needles, may stir a desire to explore more ancient forests.

CALIFORNIA

The Save-the-Redwoods League, the Sierra Club, and many other groups and individuals have worked to preserve old growth in California for more than one hundred years. Because of this California has some of the most accessible forests in the country. While other states have spectacular old-growth stands in upriver valleys or in remote, unpopulated areas, in Northern California one can visit an ancient forest fairly easily. Although some of the best may have been protected, less than 3 percent of the overall redwood forest remains. Douglas-fir forests (in some places a component of the redwood forest) are equally rare but more remote. The bitter fight to save big trees continues today, with one recent protracted battle partially concluded when the state and federal governments pur-

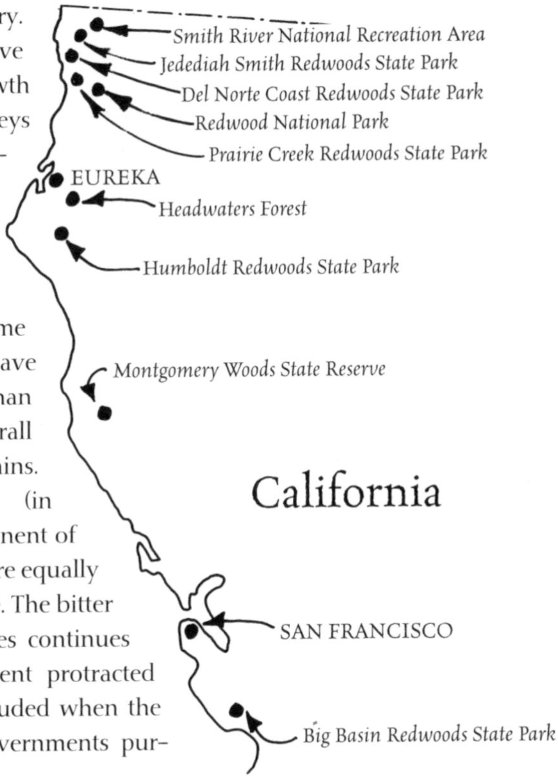

Smith River National Recreation Area
Jedediah Smith Redwoods State Park
Del Norte Coast Redwoods State Park
Redwood National Park
Prairie Creek Redwoods State Park
EUREKA
Headwaters Forest
Humboldt Redwoods State Park
Montgomery Woods State Reserve
California
SAN FRANCISCO
Big Basin Redwoods State Park

chased Headwaters Forest redwoods for $480 million, four times the highest government estimate of the land's worth. Although this seems a tidy sum, Redwood National Park's total bill came to more than $1 billion, more than the cost of all the other National Parks in the United States combined. In each case prices have been inflated because the old-growth redwoods involved were a "priceless" commodity, an irreplaceable piece of our heritage: it is a seller's market. Never mind that rather greedy and unscrupulous men became rich with the deed. Our grandchildren will undoubtedly think these purchases were at bargain basement prices!

BIG BASIN REDWOODS STATE PARK, just north of Santa Cruz, was established in 1902 and is the oldest California coast redwood preserve. With more than 18,000 acres, it is home to some of the finest redwoods south of San Francisco. At the southern extent of this great northwest forest ecosystem, Big Basin is a contrast of hot summers and mild winters, with summer fog moderating the climate enough to hold it all together. Thirty-five miles of trails through redwood groves and along ridgetops and valleys offer a splendid view of the old-growth forest at its southern limit.

MONTGOMERY WOODS STATE RESERVE, a tiny 1,300-acre redwood preserve, deserves to be mentioned because this is where the newly designated tallest tree in the world stands. Recently measured, the Mendocino Tree tops out at 367.5 feet. This listing is really a curse, however, because such fame only brings human pressures to bear that the redwood's shallow roots simply cannot tolerate. It should be enough to know that this tree exists; a visit should not be necessary.

HUMBOLDT REDWOODS STATE PARK hugs the Eel River and, at more than 60,000 acres, is California's largest redwood state park. Included here is the 9,000-acre Rockefeller Forest, considered the largest remaining chunk of old-growth redwood forest. About one in five of the remaining old-growth redwoods in the country are in this 17,000-acre park. Dyerville Giant, determined at one time to be the tallest tree on earth, fell in 1994 and is more impressive prone than it was standing. According to the Tall Trees Club, unofficially, Paradox Tree is the third tallest redwood at 366.3 feet, and nearby Rockefeller Tree is 365.3 feet.

HEADWATERS FOREST, south of Eureka, is undeveloped and currently almost inaccessible to the public, but it is the most recently preserved large chunk of old-growth redwood forest. After many years of bitter controversy, both the federal government and California state government finally purchased Headwaters Forest, preserving about 4,000 acres of old-growth redwoods included within a 10,000-acre expanse that is protected from future logging.

REDWOOD NATIONAL AND STATE PARKS include a patchwork of pristine forest and heavily cut-over lands that are slowly recuperating, as well as part of the shoreline area and the ocean offshore. It has been designated by the United Nations a World Heritage Site and International Biosphere Reserve.

The three redwood state parks listed below were purchased in the 1920s and are among the finest old-growth forests left in Northern California. Redwood National Park was established in 1968, then expanded in 1978 after more debate and conflict between the timber industry, its employees, and dedicated preservationists. Today, the three combined state redwood parks and surrounding Redwood National Park total more than 110,000 acres, but of this old-growth forest totals only about 19,640 acres. Approximately one in four of all remaining old-growth redwood trees are in these four parks.

PRAIRIE CREEK REDWOODS STATE PARK, just north of Orick, is considered by many to be the best of the best. The scenic quality of this forest, influenced by summer fog yet partially shielded from it by a hill, is unmatched. No wonder the Steven Spielberg film *Lost World* was shot here, for it truly seems as if a dinosaur could lurk behind each tree. No other old-growth forest seems to have as much organic matter as this one. With 12,500 acres, it also has the largest Roosevelt elk herd in California. Fern Canyon, an emerald-walled box canyon, supports countless lush ferns. The Sir Isaac Newton Tree, the current champion redwood at 300 feet tall and 31,670 cubic feet (the largest dimension overall), is only one of many extraordinary trees found here.

DEL NORTE COAST REDWOODS STATE PARK, just south of Crescent City, is another of the seven largest remaining California old-growth redwood forests. Sitting astride the ocean and atop a very steep ridge,

Highway 101 passes through most of the park, providing exceptional views of a roadside loaded with Pacific rhododendron and tiger lilies. Damnation Creek Trail with its switchbacks from highway to beach, provides an excellent view of the transition between the ridgetop redwoods, which don't tolerate salt-filled air, and the Sitka spruce, which love ocean shores.

JEDEDIAH SMITH REDWOODS STATE PARK, just east of Crescent City, has remarkable old growth plus the Smith River, the last large free-flowing river in California. This 10,000-acre park, accessed by both Highway 199 and the gravel Howland Hills Road, is prime redwood country, with an interesting mixture of coastal Sitka spruce, grand fir, Douglas-fir, and Port Orford cedar. This is near the north end of redwood country, and the transition to coastal Sitka spruce–western hemlock forest is interesting to investigate.

Several large and rather famous trees stand in this park, including the so-called Lost Monarch and Del Norte Titan, massive trees with reportedly the largest volumes of any redwood alive. Both are more than 300 feet tall and have diameters at 4½ feet above the ground of more than 23 feet. Both trees have estimated trunk volumes of more than 32,000 cubic feet. Lost Monarch has a volume of 35,750 cubic feet, or 440,000 board feet of wood (one board foot is 12 inches x 12 inches x 1 inch). Simply stated, this single tree has enough wood to build fifty average houses.

SMITH RIVER NATIONAL RECREATION AREA, east of Jedediah Smith Redwoods State Park, offers the pure and free-flowing Smith River, which drops from the high Siskiyou Mountains to the coast. Here, odd areas of peridotite and serpentine rock create nutrient-poor soils that inhibit "normal" growth and tall trees. Called by some the "Smoky Mountains of the West," this area is one of the oldest and most diverse botanical landscapes and contains some of the most distinctive forests in the country. Unusual soils have created miniature, ancient trees and odd insect-eating plants. That's not to say that the Smith River area of Six Rivers National Forest doesn't have its share of giant conifers; large stands of magnificent timber are easily seen from Highway 199, the main access to this area. Port Orford cedar, coast redwood, and Douglas-fir all stand tall here, towering over the river, a national treasure in its own right.

OREGON

Less than 10 percent (about 300,000 acres) of Oregon's original forest remains. Almost all of it is on public land and most is in rather remote mountain locations, so viewing is not easy. Partially because of the lack of large national parks such as those in Washington and partially because of the lack of an early conservation group such as California's Save-the-Redwoods League and the Sierra Club, Oregon was relatively late in preserving the grand forests it once grew. Still, many handsome trees and miles of old-growth forest trails await, although hikers will see clearcuts in the distance.

OREGON CAVES NATIONAL MONUMENT, about 20 miles southeast of Cave Junction, has some fine old growth with a trail that leads to what many believe is Oregon's largest Douglas-fir. This giant, about halfway along the 3.3-mile Big Tree Loop Trail, has a diameter of more than 13 feet and is estimated to be between twelve hundred and fifteen hundred years old.

UNION CREEK is a part of **ROGUE RIVER NATIONAL FOREST**, which has been heavily logged, but this area—just southwest of Crater Lake National Park on Highway 62—is an exceptional forest. Union Creek Trail 1035 is south of Union Creek Resort and located on the highway's east side. It follows the creek for 4.4 forested miles beside ancient Douglas-fir and waterfalls. In fact, many miles along the highway here run through a narrow high-quality corridor of old growth.

FRENCH PETE CREEK and the **THREE SISTERS WILDERNESS**, together designated a World Biosphere Reserve by the United Nations, are one of the few places in the Oregon Cascades where fishers (*Martes pennanti*) and wolverines may still exist. Three Sisters Wilderness spans life zones from low-elevation old growth to alpine areas. From the town of Blue River on Highway 126, turn south on Forest Service Road 19 for 11 miles to French Pete Campground. French Pete Trail 3311 is a good introduction to this spectacular area.

DELTA GROVE, in the Willamette National Forest, is home to large incense cedars and other conifers, that offer visitors a pleasant old growth walk along the river. Take Highway 126 four miles east of the town of Blue River, then south on Forest Service Road 19. Cross the McKenzie River and turn right on Forest Service Road 400 to the Delta Campground.

H. J. ANDREWS EXPERIMENTAL FOREST in the Willamette National Forest has some of the best remaining old growth in Oregon, but it is also worked harder than most. In recent years, it has been the center of great conservation battles. The Experimental Forest is north of the town of McKenzie Bridge on Highway 126 and is famous for large concentrations of western red cedar in exceptional old growth. Many studies of forest dynamics have been conducted here in a continued effort to understand how ancient forests "work."

OPAL CREEK in Willamette National Forest, the largest uncut watershed in Oregon, is one of the finest old-growth forests in the Cascades. Opal Creek Wilderness, at 13,000 acres, is connected to 35,000-acre Bull-of-the-Woods Wilderness. Opal Creek is 23 miles east of Salem on Highway 22, then left on Little North Fork Santiam Road to Forest Service Road 2299, then 5 miles on to the Opal Creek gate.

LOST LAKE, in **MOUNT HOOD NATIONAL FOREST**, is about 27 miles southwest of the town of Hood River and is probably the most impressive old growth remaining in this national forest. Several trails in the vicinity of Lost Lake pass beneath truly gorgeous ancient red-cedars and fir. Check with the Hood River ranger district office for details.

WASHINGTON

Washington's forest history is both virtuous and disastrous. Lowland ancient forests were so easily accessible from the water highways of Puget Sound, Grays Harbor, and Willapa Bay that most were cut well before 1900. Today those huge tidewater trees can only be seen in photos; what a forest this must have been! Mountains ringing the Sound were spectacular, and large numbers of trees were preserved when the Mount Rainier, Olympic, and North Cascades National Parks were created. Within several hours of driving from Washington's largest cities, spectacular and easily reachable forests await.

With almost one million acres of forests, glaciers, and beaches, **OLYMPIC NATIONAL PARK** is one of the world's great nature parks. It is famous for its old-growth forests, and some are the most heavily visited in the Northwest. The Park Service has wisely channeled visitation to a few specific areas, leaving most forests lightly visited. A walk on any trail in Olympic old growth is truly a majestic experience.

From Hurrincane Ridge, one can easily see more old growth than most any other location

Crowded but worth every step, the justly famed **HOH RAIN FOREST** offers several trails radiating from the visitors center, including a paved handicapped-access trail. The Hall of Mosses Trail meanders three-quarters of a mile past maples draped with moss; the Spruce Nature Trail loops 1¼ miles down succeeding river benches to the Hoh River and reveals how rivers help evolve a lowland river-bench forest ecosystem.

South of the Hoh, the **QUEETS AND QUINAULT VALLEYS** also have impressive old growth. Queets has the park's largest Douglas-fir, and the valley holds a narrow corridor of park old growth surrounded by a sea of clearcuts on private and Forest Service lands. Quinault Valley has more to offer, with a loop road passing through Forest Service lands to a river crossing, then returning to Highway 101 via the north shore of Quinault Lake. Several short loop trails investigate pristine forest here, one on the south shore area at Graves Creek near the road's end and another at Kestner Creek, near the northeast end of Quinault Creek. These trails pass beneath impressive western hemlock, western red cedar, Douglas-fir and big-leaf maple.

STORM KING AT LAKE CRESCENT offers an easy 1-mile trail to Marymere Falls, passing through some wonderful old growth and offering a hint of other hidden forests throughout the Olympics. Moss-draped red cedars and lacy hemlocks join some truly enormous Douglas-firs and add to the views of delicate Marymere Falls.

STAIRCASE RAPIDS LOOP TRAIL in the **LAKE CUSHMAN AREA**, the southeast corner of Olympic National Park, is only slightly drier than the rainy southwestern and western sides. In other words, while the Hoh visitors center sees 142 inches of precipitation each year, Staircase gets more than 100 inches. In both places this enormous amount of rain has grown large trees. This trail begins with a short side jaunt to a western red cedar 14 feet in diameter and continues upriver to a bridge and back, for a total loop of about 2 miles. Continuing up the river, the trail progresses directly into one of America's greatest wildernesses.

Follow Highway 165 about 6 miles past Wilkeson, then turn right on Carbon River Road another 8 miles to the **MOUNT RAINIER NATIONAL PARK** boundary. Under the huge moisture-blocking mass of

Mount Rainier, the **CARBON RIVER FOREST** in Mount Rainier National Park is one of the most easterly of the "temperate rain forests," created by the topography and by the spectacular western hemlock, red cedar, and Douglas-fir.

GROVE OF THE PATRIARCHS, also in Mount Rainier National Park, is located at the Ohanapecosh River Valley, near Stevens Canyon Entrance. The ancient trees here are upwards of a thousand years old flourish in a somewhat unique environmental situation. The huge red cedars and other forest giants here grow on a midriver island, long isolated from fire and having received abundant groundwater. The results are grand trees in a riverbottom setting. Don't think this small interpretative trail is all the old growth in the vicinity. Miles of forests line nearby highways, offering many chances for an ancient forest visual experience.

Many Mount Rainier visitors pass **NISQUALLY ENTRANCE** gate, then speed on to Paradise and the alpine meadows. Between the gate and Longmire, miles of old-growth forests offer many places to experience unaltered ancient forests. Especially interesting is Kautz Creek, where a massive mudslide and debris flow occurred in 1947. In the opening it made the process of natural forest succession and the quick recolonization that trees can make to a devastated area can be clearly seen. Imagine this area that was stark mud and rock only fifty years ago as what most lowland areas of Puget Sound looked like just eight thousand to nine thousand years ago when tidewater glaciers receded.

FEDERATION FOREST STATE PARK, along Highway 410 on the way to Mount Rainier, offers several nice old-growth forest trails looping out from an information center. West Trail wanders beneath large Douglas-fir and red cedar in what is called the Land of the Giants. This forest stands today thanks to the determined efforts of the Washington State Federation of Women's Clubs, which worked for forest preservation in the 1930s. Humboldt Redwoods State Park in Northern California also owes its existence to the California branch of this organization.

LONG ISLAND IN WILLAPA BAY, a small and narrow chunk of land, is isolated from the mainland by the estuarine tidal flats of Grays Harbor. Insulated by water, and sheltered by wet summer fogs, here stands a forest that has possibly not suffered a devastating fire for a thousand years, perhaps several thousands of years. The results clearly show a forest in harmony. Now owned by the U.S. Fish and Wildlife Service, the island is part of the Willapa Bay National Wildlife Refuge, accessible only by boat at high tide. Most of Long Island's trees have fallen to the ax, but 274 acres of protected ancient western red cedar and western hemlock still stand.

BRITISH COLUMBIA

Douglas-fir arrived within its current range only seven thousand years ago in the coastal areas of eastern Vancouver Island, where it approaches its most northerly limit. The western red cedar reached northwestern Washington and southern British Columbia about seven thousand years ago also. The Queen Charlotte Islands, where it is harder for seeds to migrate to, saw red cedar only about a thousand years ago. The Charlotte's great trees could be first-generation forest! However, ancient forests haven't fared any better in Canada: today, about 99 percent of lowland old-growth Douglas-fir are gone from British Columbia.

Several areas of British Columbia lowland old-growth forests are protected, but many areas remain threatened. Most are not easily accessible by anything but dirt roads or boat, but these forests are worth the adventure of finding them. Even if these spots are never visited in person, it should be satisfaction enough to realize that they still exist for future generations of murrelets, salamanders, and people to enjoy.

MACMILLAN PROVINCIAL PARK (Cathedral Grove), a small but popular roadside old-growth grove, sits beside Highway 4 between Parksville and Port Alberni on Vancouver Island. Although it is barely large enough to be called a forest, the park functions as a place to begin to learn what a red cedar and western hemlock ancient forest is about.

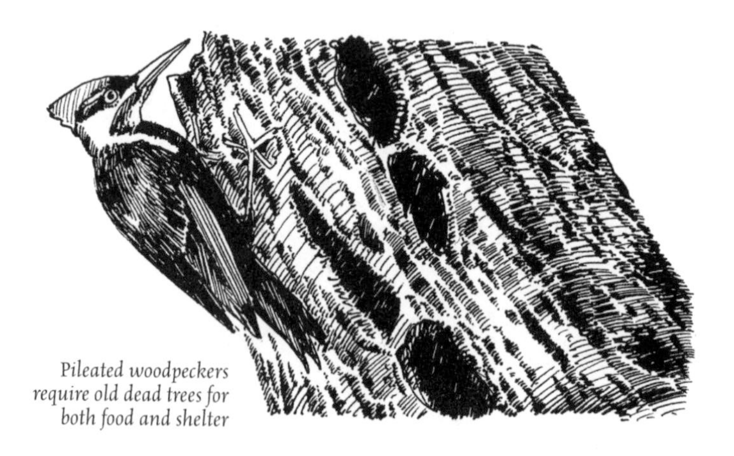

Pileated woodpeckers require old dead trees for both food and shelter

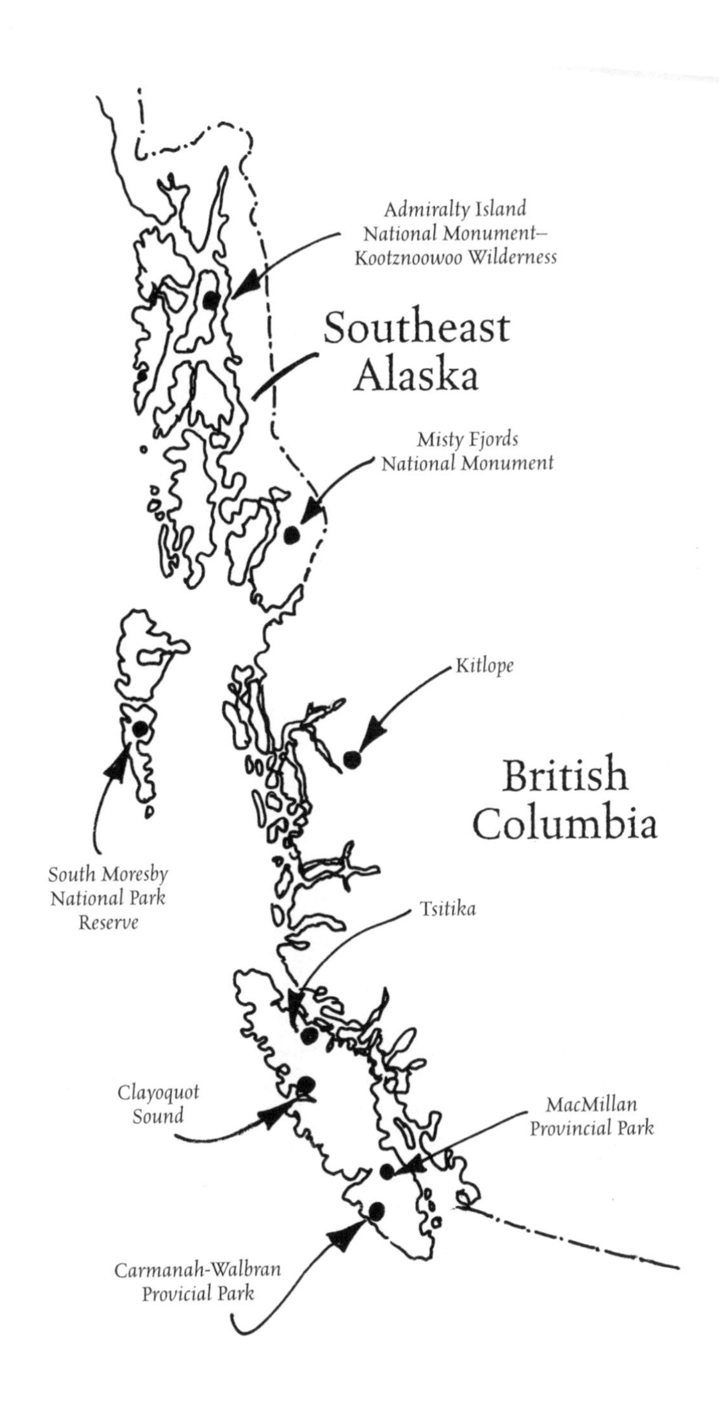
Admiralty Island
National Monument–
Kootznoowoo Wilderness
Southeast
Alaska
Misty Fjords
National Monument
Kitlope
British
Columbia
South Moresby
National Park
Reserve
Tsitika
Clayoquot
Sound
MacMillan
Provincial Park
Carmanah-Walbran
Provicial Park

CARMANAH WALBRAN PROVINCIAL PARK, once a battleground in the fight to preserve some of the biggest remaining Sitka spruce and red cedar, contains more than 40,000 acres of spectacular lowland ancient spruce forest. Located on Vancouver Island south of Bamfield, the easiest access is by turning west off Highway 1 at Duncan and then Lake Cowichan Highway 18, to the west end of Cowichan Lake. Take Rosander Main logging road (gravel with some steep grades) futher west for about two hours. Guidebooks from the 1980s don't even mention this world-class forest, but today Carmanah is home to many research scientists, and its trees have been seen in the pages of *National Geographic.*

CLAYOQUOT SOUND (pronounced *"klak-wit"*) is one of the indents on Vancouver Island's west side, just north of Tofino and Pacific Rim National Park. After years of large civil protests and countless arrests, 647,402 acres of steep-sloped fjords have been permanently placed off limits to future logging. Clayoquot can be accessed from Tofino, by both by boat and on foot.

Lowland Douglas-fir is difficult to find in British Columbia; about 99 percent of it is already gone. Thanks to orca whales (a species of threatened status), **TSITIKA RIVER** estuary and about 3,000 acres of some of the most northerly coastal Douglas-fir have been preserved. Located in Robson Bight, a slight indentation along arrow-straight Johnstone Strait just west of Kelsey Bay, this place has long been famous for the orcas that visit close to shore.

At 800,000 acres, **KITLOPE CONSERVANCY AREA** represents the finest intact old-growth ecosystem remaining on the West Coast. Remoteness has preserved this temperate rain forest, as well as the care and commitment of an Indian tribe that feels strongly that spiritual benefits of old-growth forests are more important than quick cash. Located at the head of Gardner Canal, this long and remote inland fjord, southeast of Prince Rupert, was declared preserved in 1994.

SOUTH MORESBY NATIONAL PARK RESERVE, located on the southernmost of the two largest Queen Charlotte Islands and smaller adjacent areas, encompasses a 358,000-acre preserve of spectacular forest and ocean landscape of world-class proportions. The local Haidas call it Gwaii Haanas, or "Place of Wonder," and for good reason. Preserved after many blockades, legal battles, and citizen actions across Canada, South Moresby will remain a complete ecosystem.

SOUTHEAST ALASKA

Pacific Northwest old-growth forests reach their northerly conclusion in Southeast Alaska. There's a saying that the farther out one goes, the worse it gets. This is true enough for old growth, for Alaska's hemlock-cedar forests have been whacked on for years under what many call the poorest timber policies on the West Coast. Even so, there are trees here to rejoice over. Old-growth forests are smaller here because of shorter growing seasons, but this is still a complex and unique ecosystem like its southern relatives.

MISTY FJORDS NATIONAL MONUMENT, rugged and remote, and with 3,570 square miles of steep-walled fjords, snowy mountains, and uncut forests, hugs the bottom corner of Southeast Alaska. Ketchikan is the closest city, with access to the ancient forests provided by floatplanes or boat. Comparisons with Yosemite Valley have been made of this place—for its wildness if not numbers of visitors.

ADMIRALTY ISLAND NATIONAL MONUMENT AND THE KOOTZNOOWOO WILDERNESS is a United Nations World Biosphere Reserve. The 100-mile-long island south of Juneau is mostly untouched old growth, but hand-logging scars of the past are still present in the tidewater valleys. Centrally located and accessible from Juneau, Sitka, and Petersburg, this is southeast Alaska's most visited and studied old-growth preserve.

How to Help

Simply buying this field guide contributes to the preservation of old-growth forests: a percentage of the sales of this publication are being donated to California's Save-the-Redwoods League, to be used to purchase additional redwood lands. Since 1918, the Save-the-Redwoods League has been preserving the redwood forests of California. They do this the simplest way possible: They buy redwood forest lands. Then they turn the land over to one of the many redwood parks or public reserves. To date, the League has saved more than $5 billion worth of trees and has done most of that in a positive, nonconfrontational manner.

The following organizations are also involved with expanding the public's awareness of and preserving old-growth forests. Most organizations specialize in one issue or a single area; all focus on the plight of old-growth forests and are worthy of your help, be it with contributions of money or time.

ALASKA RAINFOREST CAMPAIGN
406 G Street, Suite 209
Anchorage, AK 99501
(907) 222-2552
http://akrain.org/
The campaign works to save southeast Alaska's old growth.

ENVIRONMENTAL PROTECTION INFORMATION CENTER
351 Sprowl Creek Road
Garberville, CA 95542
(707) 923-2931
www.wildcalifornia.org
EPIC helped save Northern California's Headwaters Forest.

GOLDEN GATE NATIONAL PARK ASSOCIATION
Fort Mason Building 201, Third Floor
San Francisco, CA 94123
(415) 561-3000
The association assists Muir Woods National Monument.

HUMBOLDT REDWOODS INTERPRETATIVE ASSOCIATION
Box 100
Weott, CA 95571
(707) 946-2409
The association assists Humboldt Redwoods State Park.

MOUNTAIN PARKS FOUNDATION
525 North Big Trees Park Road
Felton, CA 95018
(831) 335-3174
The foundation assists Big Basin and Henry Cowell Redwoods State Parks.

NORTHCOAST ENVIRONMENTAL CENTER
879 Ninth Street
Arcata, CA 95521
(707) 822-6918
http://www.necandeconews.to/
The center saves old growth in Northern California and southern Oregon.

NORTH COAST REDWOODS INTERPRETATIVE ASSOCIATION
Orick, CA 95555
(707) 464-6101, ext. 5300
The association assists Prairie Creek, Del Norte Coastal, and Jedediah Smith Redwoods State Parks.

NORTHWEST INTERPRETATIVE ASSOCIATION
909 First Avenue, Suite 630
Seattle, WA 98104
(206) 220-4140
The association fosters interpretation of Olympic, Mount Rainier, and North Cascades National Parks, and Mount St. Helens National Monument, as well as other old-growth areas of the Pacific Northwest.

OREGON NATURAL RESOURCES COUNCIL
5825 N. Greely
Portland, OR 97217
http://www.onrc.org/
This coalition of fifty organizations in Oregon works to preserve old-growth forest.

REDWOOD NATURAL HISTORY ASSOCIATION
1111 Second Street
Crescent City, CA 95531
(707) 464-6101
The association assists Redwood National Park.

SAVE-THE-REDWOODS LEAGUE
114 Sansome Street, Room 605
San Francisco, CA 94104-3814
(415) 362-2352
http://www.savetheredwoods.org/
The League has been saving the redwoods since 1918.

SIERRA CLUB
85 Second Street, Second Floor
San Francisco, CA 94105-3441
(415) 977-5500
http://www.sierraclub.org/
Founded by John Muir, this organization remains instrumental in saving trees.

WESTERN CANADA WILDERNESS COMMITTEE
227 Abbott Street
Vancouver, BC, Canada V6B-2K7
(604) 683-8220 or 1-800-661-WILD
http://wildernesscommittee.org/
70,000 donors and members save old-growth in Canada.

WILDERNESS SOCIETY PACIFIC NORTHWEST
1424 Fourth Avenue, Suite 816
Seattle, WA 98101-2217
(206) 624-6430
http://www.wilderness.org/ccc/pacificnw/
This is the regional office of the national organization.

INDEX